We Can Do It!

Year 2

USING AND APPLYING MATHS CHALLENGES

Peter Clarke

Acknowledgement

The author wishes to thank Mike Askew, Sheila Ebbutt and Brian Molyneaux for their valuable contributions to this publication.

Thanks also to the following:
Karen Holman, Paddox Primary School, Rugby
Hilary Head, Send C of E First School, Surrey
Shirley Mulroy, Leslie Rankin, Hady Primary School, Chesterfield
Catherine Aket, Whitecrest Primary School, West Midlands
Kim Varden, Broke Hall Primary School, Suffolk
Elaine Richardson, St Augustine's Primary School, Cheshire
Carolyn Wallis, St Nicholas House Junior School, Hertfordshire
Lyn Wickham, Emma Bailey, Bidbury Junior School, Hampshire
Mandy Patterson, Temple Mill Primary School, Kent
John Ellard, Kingsley Primary School, Northampton
Father Rudolf Loewenstein, St Christina's Primary School, Camden
Joyce Lydford, Balgowan Primary School, Kent
Louise Guthrie, Angela Beall, Bardsey Primary School, Leeds
Sharon Thomas, Cwmbwrla Primary School, Swansea
Lynwen Barnsley, Education Effectiveness, Swansea
Jennie Jump, Advisor, Leeds
Sharon Sutton, University of Reading
Steve Lumb, Fielding Primary School, Ealing
Deborah de Gray, West Kingsdown C of E Primary School, Kent
Kerry Ann Darlington, Ullapool Primary School, Ross-shire
Helen Elis Jones, University of Wales, Bangor
Jayne Featherstone, Elton Community Primary School, Lancashire
Jane Airey, Frith Manor Primary School, Barnet
Andrea Trigg, Felbridge Primary School, West Sussex
Helen Andrews, Blue Coat School, Birmingham
Joyce Atkinson, Croham Hurst Junior School, Surrey
Jane Holmes, Elizabeth Wyles, St John's Primary School, Oxon

Thanks also to the BEAM Development Group:
Mich Bahn, Canonbury Primary School, Islington
Joanne Barrett, Rotherfield Primary School, Islington
Catherine Horton, St Jude and St Paul's School, Islington
Simone de Juan, Prior Weston Primary School, Islington

Published by BEAM Education
Nelson Thornes
Delta Place, 27 Bath Road
Cheltenham GL53 7TH

Telephone 01242 267287
Fax 01242 253695
Email cservices@nelsonthornes.com
www.beam.co.uk

ISBN 978 1 9062 2446 2
British Library Cataloguing-in-Publication Data
Data available
Edited by Marion Dill
Design by Reena Kataria
Layout by Matt Carr
Illustrations by Matt Carr
Cover photo: Lauriston School, Hackney
Printed in China by 1010 Printing International Ltd

Contents

Introduction

What is AT1: Using and applying mathematics?

Mathematical problem solving involves using previously acquired mathematical understanding, knowledge and skills and applying these to solve problems arising within everyday life, as well as within mathematics.

A major reason for studying mathematics is to be able to develop problem-solving, reasoning and logical skills that we can apply to everyday situations. We want children to employ their 'pure' mathematical knowledge effectively in real-life situations day by day, both within and outside school. It is important for children to see how acquiring mathematical understanding can help them solve problems that are relevant to their daily lives. Mathematics can help them interpret and analyse real-life situations. It can also be a source of creative pleasure for its own sake.

Using and applying mathematics is often mistaken as simply solving word problems.

Tomas shared 20 marbles equally among himself and his four friends. How many marbles did each child get?

To solve this word problem, you need to be able to read and understand what the problem is, identify the calculation you need to do, do the calculation and then interpret the answer you get in the context of the problem. In a limited sense, you are using and applying mathematics, but at a low level. Word problems at this level follow a predictable pattern, which removes the need for any real problem solving.

Of course, word problems can be more complex, and children have to work out which information is relevant, and what the context tells you about the kind of answer you need (and knowing key words is not always a helpful strategy). This multi-step problem is more like the kinds of problems that children face in real life.

Tomas is 8 years old, and he gets 24 marbles. He keeps half of the marbles for himself and shares out the rest equally among his four friends. How many more marbles does he have than each of his friends?

The point about more complex problems is that you have to work out what the meaning is, what sort of outcome you need, and what sensible calculations to do to get there. Problems in real life are mainly like this.

An investigative approach to the simple marble problem could be:

How many different ways can you share 20 marbles equally?

A more complex investigation could be:

It's Tomas's birthday, so he gets more marbles than anyone else. Everyone else gets the same amount. How many different ways can you share the 24 marbles?

Sometimes it is interesting just exploring some mathematics for its own sake, as a pastime. Some of the problems in **We Can Do It!** are like that. The interest lies in using your brain, finding a pattern, seeking a neat solution – just like doing a crossword or sudoku puzzle. In the process of working on the investigation, children will be honing their reasoning skills and using their creativity to seek a way forward.

Some problems really do mirror everyday life:

If we spend less time tidying up, will we get more time for playing outside?

What news stories on the front page of the newspaper are given the most space?

What proportion of water do you need to dilute a fruit drink?

Many of these involve measures and data handling. Children need practical experiences to solve these: if you don't know what 100 ml of liquid looks like, and how it compares with 1 litre, you cannot solve the problem in the abstract. This means being prepared for a busy, active and, at times, messy classroom – and also a classroom where children discuss with each other the problem in hand.

Real and complex problems and investigations require children to search for strategies to get started and to draw upon their experiences and knowledge of 'pure' mathematics. They also encourage children to work flexibly, creatively and logically. They are less comfortable for the teacher because the outcomes are not always predictable and the answers are not always known. Our role is to work with children, sometimes doing the mathematics alongside them, looking for and encouraging creative and logical thinking, rather than focusing on right answers.

Mathematical thinking

The *National Curriculum* (2000) outlines the thinking skills that complement the key understanding, knowledge and skills that are embedded in the statutory primary curriculum.

The *We Can Do It!* series aims to develop the following key thinking skills in children:

Information – processing skills

- Locate, collect relevant information
- Sort, classify, sequence, compare and analyse part and/or whole relationships

Reasoning skills

- Give reasons for opinions and actions
- Draw inferences and make deductions
- Use precise language to explain what they think
- Make judgements and decisions informed by reason or evidence

Enquiry skills

- Ask relevant questions
- Pose and define problems
- Plan what to do and how to research
- Predict outcomes and anticipate conclusions
- Test conclusions and improve ideas

Creative thinking skills

- Generate and extend ideas
- Suggest hypotheses
- Apply imagination
- Look for alternative innovative outcomes

Evaluative skills

- Evaluate information
- Judge the value of what they read, hear or do
- Develop criteria for judging the value of their own and others' work or ideas
- Have confidence in their judgement

We Can Do It!

In this series, we provide problems and challenges that stimulate genuine mathematical thinking. These problems are written for a community of learners in the primary classroom – that is, we expect the problems to be solved collaboratively by pairs of children, groups and whole classes working together and discussing the problems at every stage. With each problem, we offer teaching advice on how to encourage high-level thinking among children. We also analyse each problem and children's possible responses to it in order to promote greater understanding of how children develop problem-solving skills.

The challenges in *We Can Do It!* are designed to improve children's attainment in the three strands of AT1 of the *National Curriculum* (2000): Using and applying mathematics.

In **problem solving** by:

- using a range of problem solving strategies

- trying different approaches to a problem

- applying mathematics in a new context

- checking their results

In **communicating** by:

- interpreting information

- recording information systematically

- using mathematical language, symbols, notation and diagrams correctly and precisely

- presenting and interpreting methods, solutions and conclusions in the context of the problem

In **reasoning** by:

- giving clear explanations of their methods and reasoning

- investigating and making general statements

- recognising patterns in their results

- making use of a wider range of evidence to justify results through logical reasoned argument

- drawing their own conclusions

The challenges also provide children with an opportunity to practise and consolidate the five themes and objectives of Strand 1: Using and applying mathematics for Year 2 in the *Renewed Framework of Mathematics* (2006).

Creating a problem-solving classroom

It is important that children have faith in their own abilities and develop a healthy self-esteem. They need to be encouraged to have a go, even if at first their attempts are wrong. We want children to realise that having a go and making a mistake is far better than not attempting a problem at all, and that trial and improvement is a vital part of the learning process. Therefore it is important to encourage and reward the following qualities during problem-solving lessons:

- perseverance

- flexibility

- originality

- active involvement

- independence

- cooperation

- willingness to communicate and share ideas

- willingness to try and take risks

- reflection

Teacher expectations are a critical factor affecting children's achievement. We can engender a classroom ethos that makes anything possible for all children. We can offer children opportunities to reach their full potential, regardless of supposed appropriate year-level expectations.

Assessment

You can use the challenges in *We Can Do It!* with the whole class or with groups of children as an assessment activity. Linked to the strand that is being studied at present, *We Can Do It!* will not only provide you with an indication of how well the children have understood the 'pure' mathematics objectives being covered, but also their problem-solving skills.

Throughout each of the challenges there are prompting questions which focus on specific aspects of the challenge. At the end of each challenge there are also three questions that are specifically designed to help with assessing using and applying mathematics.

The list of thinking-skills statements on page 7 and the descriptions relating to the three strands of AT1 on page 8 are extremely useful in helping assess children's problem-solving skills.

Problem-solving strategies

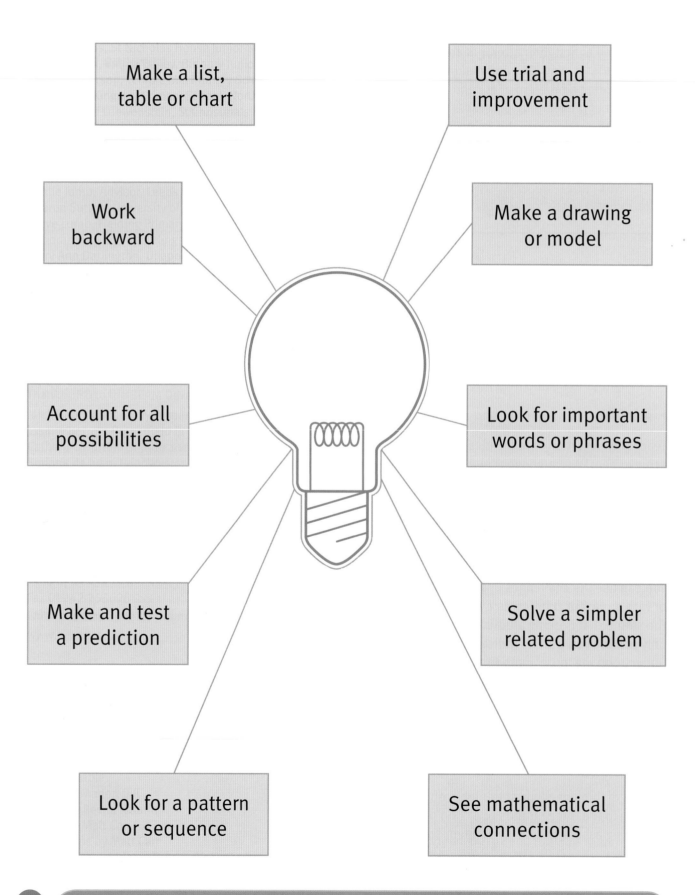

Make a list, table or chart

Use trial and improvement

Work backward

Make a drawing or model

Account for all possibilities

Look for important words or phrases

Make and test a prediction

Solve a simpler related problem

Look for a pattern or sequence

See mathematical connections

A final word

From an early age, children can learn that school mathematics is 'work' – a series of tasks they need to get through as quickly as possible, preferably without the need for thought. The challenges in **We Can Do It!** are deliberately demanding for children in order to promote their ability to solve problems. You will need to encourage them to rely less on your help, setting them off to work on a challenge for a short length of time. Follow this with time together to discuss the different ways in which they have set about the task; this will help them realise that they can achieve something, while you feed in ideas for continuing without taking the responsibility for the thinking away from the children.

Being challenged is enjoyable! The challenges in **We Can Do It!** have not been 'dressed up' to disguise the mathematics or to make them 'fun'. The aim is not to make mathematics itself enjoyable but rather find enjoyment by being prepared to have a go at something, rising to the challenge and reaching a satisfactory conclusion.

How to use this book

Question to pose the challenge

Opening question to ask the children that is designed to act as a springboard into the challenge

Summary of maths content

Brief summary of the 'pure' mathematics focus of the challenge

Introducing the challenge

Outline scenario to hook in the children's interest. It often includes opportunities to engage the children's interest further by including 'turn and talk' instructions.

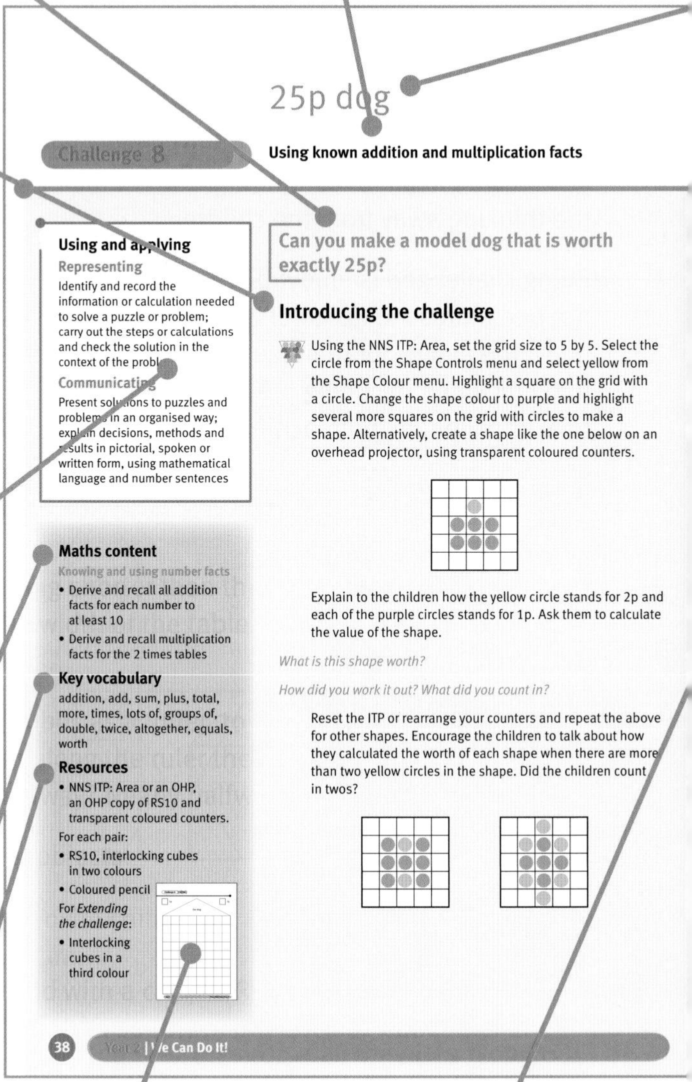

Using and applying

Description of how the challenge links to the five themes in Strand 1: Using and applying mathematics), in the *Renewed Framework for Mathematics* (2006)

Solving problems

Representing

Enquiring

Reasoning

Communicating

Maths content

Objectives from the *Framework* specifically covered in the challenge

Key vocabulary

List of words and phrases appropriate to the challenge

Resources

List of resources children need to undertake the challenge, including resource sheets (RS)

RS diagram

You find the resource sheet (RS) on the CD.

The challenge

This offers advice on how to structure the challenge and uses the following symbols for clarification:

 individual paired

 group whole class

(Sample page content)

25p dog

Challenge 8

Using known addition and multiplication facts

Using and applying

Representing
Identify and record the information or calculation needed to solve a puzzle or problem; carry out the steps or calculations and check the solution in the context of the problem

Communicating
Present solutions to puzzles and problems in an organised way; explain decisions, methods and results in pictorial, spoken or written form, using mathematical language and number sentences

Maths content
Knowing and using number facts
- Derive and recall all addition facts for each number to at least 10
- Derive and recall multiplication facts for the 2 times tables

Key vocabulary
addition, add, sum, plus, total, more, times, lots of, groups of, double, twice, altogether, equals, worth

Resources
- NNS ITP: Area or an OHP, an OHP copy of RS10 and transparent coloured counters.
For each pair:
- RS10, interlocking cubes in two colours
- Coloured pencil
For *Extending the challenge*:
- Interlocking cubes in a third colour

Can you make a model dog that is worth exactly 25p?

Introducing the challenge

 Using the NNS ITP: Area, set the grid size to 5 by 5. Select the circle from the Shape Controls menu and select yellow from the Shape Colour menu. Highlight a square on the grid with a circle. Change the shape colour to purple and highlight several more squares on the grid with circles to make a shape. Alternatively, create a shape like the one below on an overhead projector, using transparent coloured counters.

Explain to the children how the yellow circle stands for 2p and each of the purple circles stands for 1p. Ask them to calculate the value of the shape.

What is this shape worth?

How did you work it out? What did you count in?

Reset the ITP or rearrange your counters and repeat the above for other shapes. Encourage the children to talk about how they calculated the worth of each shape when there are more than two yellow circles in the shape. Did the children count in twos?

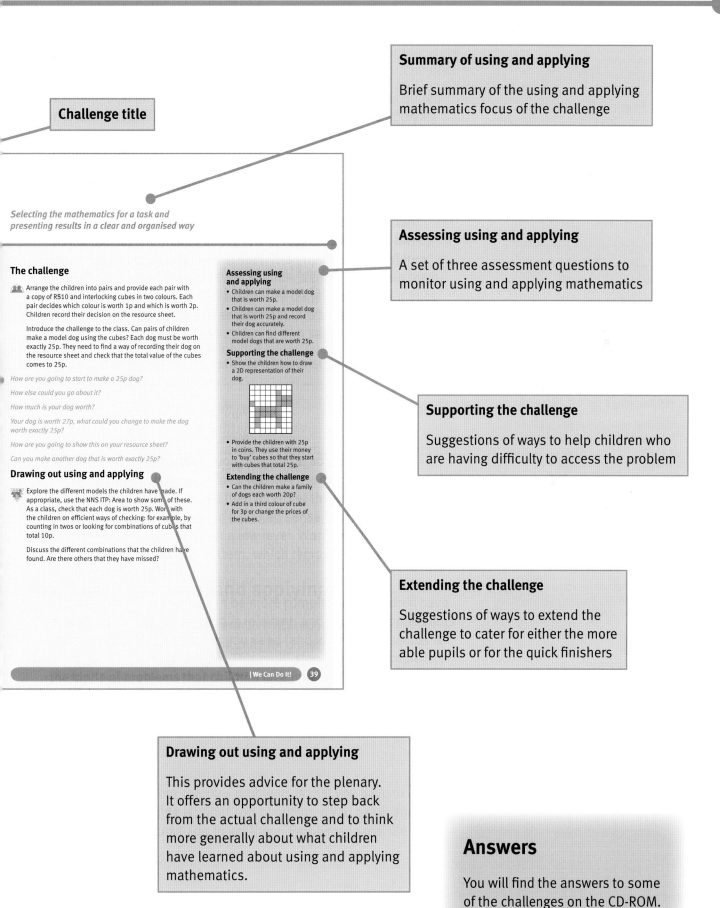

Challenge title

Summary of using and applying

Brief summary of the using and applying mathematics focus of the challenge

Selecting the mathematics for a task and presenting results in a clear and organised way

The challenge

Arrange the children into pairs and provide each pair with a copy of RS10 and interlocking cubes in two colours. Each pair decides which colour is worth 1p and which is worth 2p. Children record their decision on the resource sheet.

Introduce the challenge to the class. Can pairs of children make a model dog using the cubes? Each dog must be worth exactly 25p. They need to find a way of recording their dog on the resource sheet and check that the total value of the cubes comes to 25p.

How are you going to start to make a 25p dog?

How else could you go about it?

How much is your dog worth?

Your dog is worth 27p, what could you change to make the dog worth exactly 25p?

How are you going to show this on your resource sheet?

Can you make another dog that is worth exactly 25p?

Drawing out using and applying

Explore the different models the children have made. If appropriate, use the NNS ITP: Area to show some of these. As a class, check that each dog is worth 25p. Work with the children on efficient ways of checking: for example, by counting in twos or looking for combinations of cubes that total 10p.

Discuss the different combinations that the children have found. Are there others that they have missed?

Assessing using and applying
- Children can make a model dog that is worth 25p.
- Children can make a model dog that is worth 25p and record their dog accurately.
- Children can find different model dogs that are worth 25p.

Supporting the challenge
- Show the children how to draw a 2D representation of their dog.

- Provide the children with 25p in coins. They use their money to 'buy' cubes so that they start with cubes that total 25p.

Extending the challenge
- Can the children make a family of dogs each worth 20p?
- Add in a third colour of cube for 3p or change the prices of the cubes.

| We Can Do It! | 39

Assessing using and applying

A set of three assessment questions to monitor using and applying mathematics

Supporting the challenge

Suggestions of ways to help children who are having difficulty to access the problem

Extending the challenge

Suggestions of ways to extend the challenge to cater for either the more able pupils or for the quick finishers

Drawing out using and applying

This provides advice for the plenary. It offers an opportunity to step back from the actual challenge and to think more generally about what children have learned about using and applying mathematics.

Answers

You will find the answers to some of the challenges on the CD-ROM.

Lesson suggestions

Aspects of mathematical problem solving should be covered in every maths lesson, even those that aim to teach the purest of mathematical concepts. Children need to see the application of 'pure' maths in everyday experiences.

We advise, however, that once a week you devote a lesson entirely to developing children's problem-solving skills. It is for this reason that **We Can Do It!** consists of 36 challenges.

The challenges in this book provide children with an opportunity to practise and consolidate the Year 2 objectives from the *Renewed Framework for Mathematics* (2006). The curriculum charts on pages 18-21 show which challenge is matched to which planning block and mathematics strand. Refer to these charts when choosing a challenge.

We Can Do It! and the daily maths lesson

The challenges in *We Can Do It!* are ideally suited for the daily maths lesson. You can introduce each challenge to the whole class or to groups of children. Here is a suggestion how to structure a lesson using *We Can Do It!*.

Introducing the challenge

- Introduce the idea of the challenge either as a discussion or by giving a simplified version of the problem. You may need to highlight some of the mathematics children need to solve the problem.

- Introduce the challenge to the children by asking the question that poses the problem.

- Stimulate children's involvement through discussion.

- Use the key vocabulary throughout and explain new words where necessary.

- Make sure that the children understand the challenge.

- If you use a resource sheet, make sure children understand the text on the sheet.

- Begin to work through the challenge with the whole class, pointing out possible problem-solving strategies.

The challenge

- Arrange children into pairs or groups to work on the problem.

- Make sure appropriate resources are available to help children with the challenge.

- Monitor individuals, pairs or groups of children, offering support when and where needed.

- If appropriate, extend the challenge for some children.

Drawing out using and applying

- Plan an extended plenary.

- Discuss the challenge with the class.

- Invite individual children, pairs or groups to offer their solutions and the strategies they used.

The teacher's role in problem-solving lessons

- Give a choice where possible.

- Present the problem orally, giving maximum visual support where appropriate.

- Help children 'own the problem' by linking it to their everyday experiences.

- Encourage children to work together, sharing ideas for tackling a problem.

- Allow time and space for collaboration and consultation.

- Intervene, when asked, in such a way as to develop children's autonomy and independence.

- Work alongside children, setting an example yourself.

- Encourage the children to present their work to others.

Paired and group work

We Can Do It! recognises the importance of encouraging children to work collaboratively. All of the challenges in *We Can Do It!* include some element of paired or group work. By working as a group, children develop cooperation and collective responsibility. They also learn from each other, confirming their mathematical knowledge and identifying for themselves, in a non-threatening environment, any misconceptions they may hold.

The *National Curriculum* identifies three strands of the AT1: Using and applying mathematics. They are problem solving, communicating and reasoning. While it is possible for children to problem solve independently, communication, as the diagram below illustrates, is a cooperative, interactive process that involves both expressing and receiving information.

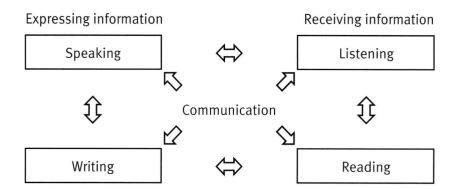

Meaningful reasoning can only occur through communication. Children cannot effectively reason with themselves: they always see themselves as being right! It is not until they begin to discuss and share ideas with others that children begin to reason and see other points of views and possibilities.

Charts linking to the
Renewed Framework for Mathematics (2006)

Chart linking blocks and strands of the *Renewed Framework for Mathematics* (2006)

	Strand 2: Counting and understanding number	Strand 3: Knowing and using number facts	Strand 4: Calculating	Strand 5: Understanding shape	Strand 6: Measuring	Strand 7: Handling data
BLOCK A: Counting, partitioning and calculating	●	●	●			
BLOCK B: Securing number facts, understanding shapes		●		●		
BLOCK C: Handling data and measures					●	●
BLOCK D: Calculating, measuring and understanding shape			●	●	●	
BLOCK E: Securing number facts, calculating, identifying relationships	●	●	●			

Chart linking challenges in *We Can Do It! Year 2* to strands of the *Renewed Framework for Mathematics* (2006)

Challenge Number	Title	Strand 1: Using and applying mathematics — Solving problems	Representing	Enquiring	Reasoning	Communicating	Strand 2: Counting and understanding number	Strand 3: Knowing and using number facts	Strand 4: Calculating	Strand 5: Understanding shape	Strand 6: Measuring	Strand 7: Handling data
1	Dotty numbers				●	●	●					
2	Calculator counting				●	●	●					
3	Odds and evens				●	●	●	●				
4	Lots of names				●	●	●	●				
5	Handfuls halving				●	●	●					
6	Order, order!				●	●	●					
7	Pavements	●	●					●	●			
8	25p dog		●			●		●				
9	Domino stars		●			●		●				
10	Line chunks				●	●	●	●				

Number	Title	Strand 1: Using and applying mathematics					Strand 2: Counting and understanding number	Strand 3: Knowing and using number facts	Strand 4: Calculating	Strand 5: Understanding shape	Strand 6: Measuring	Strand 7: Handling data
		Solving problems	Representing	Enquiring	Reasoning	Communicating						
11	Money in my purse	●			●		●	●				
12	Making 20	●	●					●				
13	Paul's pets	●	●					●				
14	Number bracelets	●	●						●			
15	Collect the coins		●		●	●		●	●			
16	Calculating digit cards		●			●			●			
17	Number families		●		●			●	●			
18	Equal groups		●		●				●			
19	Remainder 1					●			●			
20	Name snakes				●	●		●	●			
21	Missing numbers and symbols		●			●		●	●			
22	Colour the square				●	●				●		
23	Two-piece tangram				●	●				●		

Number	Title	Strand 1: Using and applying mathematics					Strand 2: Counting and understanding number	Strand 3: Knowing and using number facts	Strand 4: Calculating	Strand 5: Understanding shape	Strand 6: Measuring	Strand 7: Handling data
		Solving problems	Representing	Enquiring	Reasoning	Communicating						
24	Make it my way				●	●				●		
25	Symmetrical patterns				●	●				●		
26	From start to finish		●			●				●		
27	Turning				●	●				●		
28	Making a ruler			●		●					●	
29	Puffed and popped			●		●					●	●
30	Which is the larger?			●		●					●	
31	Just a minute!		●			●					●	
32	Birthday month			●		●					●	●
33	Toy sort			●		●						●
34	Brothers and sisters			●		●						●
35	Birthday months			●		●						●
36	Dice totals			●		●		●				●

The challenges

Strand 2: Counting and understanding number

Strand 3: Knowing and using number facts

Strand 4: Calculating

Strand 5: Understanding shape

Strand 6: Measuring

Strand 7: Handling data

Dotty numbers

Counting large numbers by grouping

Using and applying

Reasoning

Describe patterns and relationships involving numbers and shapes; make predictions and test these with examples

Communicating

Present solutions to problems in an organised way; explain decisions, methods and results in pictorial, spoken or written form, using mathematical language

Maths content

Counting and understanding number

- Read and write two-digit numbers in figures and words
- Count up to 100 objects by grouping them and counting in twos, fives or tens

Key vocabulary

number, count, estimate

Resources

- Poster showing arrays of circles or NNS ITP: Area
- RS1 (for each child)

For *Supporting the challenge:*

- Calculator

For *Extending the challenge:*

- RS2

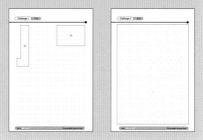

> ## How can you count the number of dots you have circled without counting every one?

Introducing the challenge

Work either with a poster showing some arrays of circles or the NNS ITP: Area. If using the latter, set the grid size to 85 by 5. Select the circle from the Shape Controls menu. Ask the children to close their eyes and quickly highlight a number of squares on the grid with a circle.

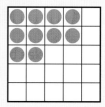

Tell the children that they are only going to get a quick look at the number of circles. They will not have time to count them all, so they need to estimate quickly how many circles there are. Ask the children to estimate the number of circles. Briefly reveal the number of circles, giving the children long enough to get a sense of how many there are, but not long enough to count them all. Call for individual responses before counting the circles.

Can you estimate how many there are without counting every one?

How did you get that estimate?

Remind the children that when making an estimate, they do not have to count each circle but should try to get a 'feel' for the number of circles.

Repeat several times before changing the grid size and asking the children to estimate larger numbers of objects.

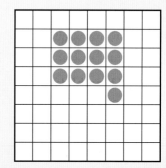 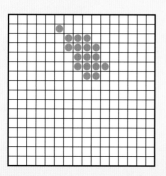

The challenge

 Still using the NNS ITP: Area, reset the ITP to the pinboard and draw a shape around some of the dots.

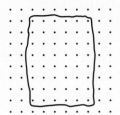

Children count exactly how many dots there are inside the shape, then explain how they counted them.

Jashan, how many dots do you think there are inside the shape? What about you, Jessica?

Reset the ITP and repeat for another shape.

 Give each child a copy of RS1. Display the resource sheet on the interactive whiteboard or as an enlarged poster. Discuss the number of dots in each shape drawn. Compare some different ways of counting the 22 and 48 dots.

Children find their own ways of grouping the dots by filling the resource sheet with shapes. Can they find the total number of dots in each shape they draw? Encourage the children to find quick ways of counting the dots they have encircled without counting every one.

Do twos help? What about tens?

 Children compare their resource sheets and discuss them.

Drawing out using and applying

 Individual children explain their favourite methods of grouping and counting the dots to the rest of the class. Using the NNS ITP: Area, explore with the children effective ways of counting larger regions of dots and efficient strategies for counting and adding groups of numbers. Explore which groups work best.

Assessing using and applying

- Children can estimate a small number of objects that can be checked by counting.
- Children can count large amounts by grouping them and counting in twos, fives and tens, writing down amounts as they go.
- Children can count large amounts mentally by grouping them and counting in twos, fives and tens.

Supporting the challenge

- Work with the children on drawing loops around sets of dots, in tens or five.
- Children work in pairs, taking turns to draw a shape round some of the dots and both working out how many dots have been encircled in each shape.

Extending the challenge

- Ask the children: "How many ways of encircling 20 dots can you find?"
- Repeat the activity with RS2 (triangular dotty paper) and compare the results.

Calculator counting

Counting, reading and writing in tens

Using and applying

Reasoning

Describe patterns and relationships involving numbers; make predictions and test these with examples

Communicating

Present solutions to puzzles and problems in an organised way; explain decisions, methods and results in pictorial, spoken or written form, using mathematical language and number sentences

Maths content

Counting and understanding number

- Read and write two-digit numbers in figures and words
- Describe and extend number sequences

Key vocabulary

number, digit, write, sequence, pattern, add, more, equals

Resources

For each pair:
- Calculator
- RS3

For *Supporting the challenge*:
- 100-grid

Can you predict the next number in the sequence?

Introducing the challenge

 Show the children how to operate the constant facility on the calculator. Different types of calculators have different rules, but one of the most common is to key in:

 Explain to the children that they are going to explore sequences when counting on in twos. Tell them to take turns to key in a number under 10 and ask their partner to count on 2 and predict the next number, or numbers, if they can. They check with the calculator.

Stop the sequences at 30. Children say the numbers as they are shown on the display, then record some sequences. Ask the children what they observe about the sequences they get.

What can you say about these numbers?

Do you recognise these numbers?

What number do you think will come next?

The challenge

 Working in pairs, children decide on a two-digit number. They record this on RS3. They set this number in their calculator, then key in the constant function to keep adding 10:

Again, they take turns to ask their partner to count on 10 and predict the next number. They check with the calculator.

They continue to take turns and make sequences to 100. Continue encouraging the children to discuss the sequence of numbers and to predict the next number in the sequence. Ask them to explain the pattern they see.

What is the next number?

*What do you think the next number will be? Why do you say that?
Were you right?*

Repeat the activity, starting with any two-digit number.

Encourage the children to present their results clearly. Remind
them to try and predict the next number in the sequence.

Drawing out using and applying

 Bring the class back together and discuss and compare the
lists of numbers. As a class, read one sequence of numbers
aloud. Children explain the patterns they see (and hear). Did
anyone reach a point in recording their patterns where they did
not need to use the calculator? Can they continue a pattern on
paper that was started on a calculator?

Encourage the children to extend the investigation by
responding to the question:

What would happen if we started with 105 or 102?

Assessing using and applying

- Children can add 10, using a calculator, and see a pattern.
- Children can recognise and explain the number sequence and use this to predict the next number in the sequence.
- Children can make a generalisation and provide examples to substantiate this.

Supporting the challenge

- Children start from 0 and add 10 each time.
- Children record the patterns on a 100-grid.

Extending the challenge

- Try starting with a two-digit or three-digit number and counting backward in twos, fives or tens.
- Try counting and recording the pattern of adding hundreds.

Odds and evens

Using known addition facts to explore number patterns

Using and applying

Reasoning

Describe patterns and relationships involving numbers; make predictions and test these with examples

Communicating

Present solutions to problems in an organised way; explain decisions, methods and results in pictorial, spoken or written form, using mathematical language and number sentences

Maths content

Counting and understanding number

- Read and write two-digit numbers in figures; recognise odd and even numbers

Knowing and using number facts

- Derive and recall all addition facts for each number to at least 10

Key vocabulary

number, addition, add, sum, total, plus, more, altogether, equals, pattern, odd, even, number sentence

Resources

- NNS ITP: 20 cards
- RS4 (for each child)
- Counters, interlocking cubes or Numicon (optional)

For *Supporting the challenge*:

- Number line

What do you notice about your answers?

Introducing the challenge

 Using the NNS ITP: 20 cards, make a stack of 20 cards, with a maximum number of 20 and a minimum number of 0. Hide the Make Stacks menu and deal the stack of cards. Divide the lower half of the screen in half and label one half 'even' and the other half 'odd'. Reveal the first card in the row: for example, 6.

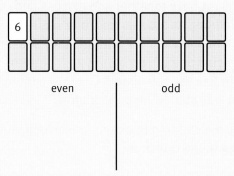

even   odd

Children say whether the number is odd or even. Once the class has agreed, drag and drop the number card into the correct set.

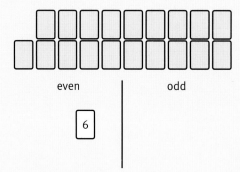

even   odd

Repeat until all the numbers have been revealed and placed in the correct set.

The challenge

 Give each child a copy of RS4. Briefly discuss the numbers at the top of the sheet. Is each number odd or even?

Children who have difficulty in identifying odd and even numbers can put out counters, interlocking cubes or Numicon corresponding to each number. Can they split the cubes into two equal sets?

Children record the numbers in the appropriate sets on the resource sheet. They then add five more numbers of their choice to each set.

 Children make an addition calculation by choosing one number from each set. They record their calculation and answer in the lower half of the resource sheet. Leave them to create several more calculations in this way.

Drawing out using and applying

 Discuss with the children what they notice about the answers to their calculations.

What do you notice about all of your answers?

What can you say about the answer to any addition number sentence that involves adding an even and an odd number together?

What do they think would happen if they chose both numbers from the set of even numbers or if they chose both from the set of odd numbers?

Set the children off to investigate these possibilities, then bring the class back together to discuss the results.

What do you notice about all of your answers?

What happens when you add two even numbers together?

What do you notice about the answer when you add two odd numbers together?

Does this always work?

Can you tell me any rules about adding odd and even numbers?

Assessing using and applying

- Children can identify odd and even numbers to at least 20.
- Children can say what they notice about the answers when they add an odd and an even number, and a pair of odd numbers and a pair of even numbers.
- Children can make a generalisation about adding different combinations of pairs of odd and even numbers: for example, O + E = O, O + O = E and E + E = E.

Supporting the challenge

- Provide children with counters, interlocking cubes or Numicon to help them identify odd and even numbers.
- Children record their answers on a number line.

Extending the challenge

- Change the operation: use subtraction or multiplication.
- Add three numbers chosen from the two sets: for example, O + E + E.

Lots of names

Counting in groups

Using and applying

Reasoning

Describe patterns and relationships involving numbers; make predictions and test these with examples

Communicating

Present solutions to puzzles and problems in an organised way; explain decisions, methods and results in pictorial, spoken or written form, using mathematical language and number sentences

Maths content

Counting and understanding number

- Count to 100 objects by grouping them and counting in tens, fives or twos

Knowing and using number facts

- Derive and recall multiplication facts and the related division facts; recognise multiples

Key vocabulary

count, groups, times

Resources

For each pair:
- Supply of RS5 (optional)
- One-minute timer or stopwatch

For *Extending the challenge*:
- Matchsticks
- Interlocking cubes

How many names and letters did you write in a minute?

Introducing the challenge

 Play 'Say it and you're out'. Ask all the children to stand behind their chair. Count on with the class in fives from 5 to 100. Repeat if necessary.

Ask the children for three of the numbers they just said (or three multiples of 5) from 5 to 100 and write these on the board: for example, 20, 55 and 90.

Explain to the children that they are going to count round the room in fives from 5 to 100. When one child says the number 100, the next child starts the count again from 5. However, every time someone says the numbers 20, 55 or 90, they are out and have to sit down.

(Variation: When one child says the number 100, the next child says 95 and the children count back in fives from 100 to 5. As the children count round the room, the count keeps going on from 5 to 100 and back from 100 to 5.)

Choose a child to start the count from 5. The child standing next to them says '10', the next child '15', and the next child '20'. The child that says '20' is out and has to sit down. The next child says '25', and so on. The winner is the last child left standing. When there are only three children left, can the class predict who will be the winner?

The challenge

 Arrange the children into pairs and provide each pair with a supply of RS5 and a one-minute timer or stopwatch. Each child takes a copy of the resource sheet and fills in the number of letters in their name at the top of the sheet.

Ask the children to decide who will be the Timekeeper first and who will be the Writer. The Timekeeper takes the one-minute timer or stopwatch. Ask the Writer to write their name on their copy of the resource sheet as many times as possible in one minute.

Children then swap roles.

Children count the number of times they wrote their name and the total number of letters, comparing results with each other. They then repeat the activity to see if they can break their record.

Drawing out using and applying

 Ask the children how they worked out how many names and how many letters.

How many times did you write your name?

How many letters is this?

How did you count the number of times you wrote your name? Was this different from how you counted how many letters you wrote altogether?

What did you do differently when you counted all the letters you wrote?

> Discuss counting in groups of 5, 10, 15 … to make counting quicker. If a child has a long name, you could suggest breaking it into smaller 'chunks' for counting.

Did you count in groups? What size groups did you count in? Why did you count in groups of that size?

How could you make it easier to count how many letters you wrote altogether?

> Discuss with the children why, when two children have written the same number of letters such as 28, one has written their name four times and the other seven times. Encourage them to look for patterns.

Yolanda and Paul have both written 28 letters, but Yolanda only wrote her name four times, whereas Paul wrote his name seven times. Why do you think that is?

What patterns can you see? Can you see any others?

Assessing using and applying

- Children can count the number of times they wrote their name in one minute and how many letters this is altogether.

- Children can work out how many letters they wrote altogether by multiplying the number of times they wrote their name by the number of letters in their name.

- Children can identify and explain why, for example, when two children have written the same number of letters such as 28, one has written four names and the other seven.

Supporting the challenge

- When counting the number of letters they wrote altogether, help the children cluster letters into groups of five.

| Angelina | Angelina | Angelina |
| Angelina | Angelina | Angelina |

- Children check each other's counting of how many letters they wrote altogether.

Extending the challenge

- Use matchsticks. One child makes triangles, the other makes squares. See how many they can each make in a minute and how many sticks they have each used.

- Use interlocking cubes. One child clips together five and the other 10. See how many sticks they make in a minute and how many cubes they have each used.

Handfuls halving

Recognising halves (and quarters) of amounts and odd and even numbers

Using and applying

Reasoning

Describe patterns and relationships involving numbers; make predictions and test these with examples

Communicating

Present solutions to puzzles and problems in an organised way; explain decisions, methods and results in pictorial, spoken or written form, using mathematical language and number sentences

Maths content

Counting and understanding number

- Recognise odd and even numbers
- Find one half (and one quarter) of sets of objects

Key vocabulary

number, half, halve, quarter, odd, even, equal

Resources

- Individual whiteboard and marker (for each pair)
- NNS ITP: Counting on and back in ones and tens or 1–100 bead string

For each child:

- Tray of counters or similar counting materials
- RS6

For *Supporting the challenge*:

- Numicon

Which numbers can you halve?

Introducing the challenge

 Arrange the children into pairs and provide each pair with an individual whiteboard and marker. Using the NNS ITP: Counting on and back in ones and tens, hide the label for the number of beads on the left side of the string and also hide the label for the number of beads the mouse pointer is on. Use the mouse to divide the line of beads into two separate smaller groups.

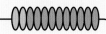

Point to the beads and ask the children to write down how many beads they think there are to the left of the gap.

How many beads do you think there are here? Quickly talk about it with your partner and write your answer on your whiteboard.

Next, ask the children to hold up their whiteboards before revealing the number of beads on the ITP or writing this up if using a bead string.

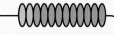

28

How did you work out that there were 28 beads?

Did anyone work it out differently?

Repeat the above several times.

Invite a child to come to the front of the class and take a handful of counters from a tray. Discuss estimates of how many counters this may be. Get the child to arrange the counters in a way that makes it easy to count them.

How many counters do you think Talma has taken?

How could Talma arrange the counters to make it easier to count exactly how many there are?

Ask the children if it is possible to halve this amount. Encourage them to discuss what this means, reminding them that each half has to be equal.

What does the word 'halve' mean?

*Do you think it is possible to halve the number of counters that
Talma took?*

Give the child at the front of the class a copy of RS6 and ask
them to use the divided shape at the top of the resource sheet
to find out if the handful divides into halves. (Note: If using
an OHP, place a wooden ruler vertically down the centre of the
OHP to divide the screen into two halves.) Discuss the results.

*Is it possible to divide the handful of counters that Talma took
in half?*

How do we know it is possible? How do we know it is impossible?

Repeat with a different handful of counters.

The challenge

 Provide each child with a copy of RS6 and a tray of counters or
similar counting materials. Children think of numbers that can
be halved and some that they think cannot. Challenge them
to find out about amounts between 10 and 20. Children with
a good knowledge of even numbers to 20 might try amounts
larger than 20: for example, between 20 and 30 or 20 and 50.
Remind the children to present their results clearly, using the
resource sheet.

Drawing out using and applying

 Children exchange their results and check each other's work.
They explain to each other how they set about solving the
challenge and justify their answers.

 Compare the children's different methods of recording and
discuss the advantages of each. Ask them if they had any quick
ways of organising the counting of the larger amounts.

How did you keep a record of the numbers you tried?

*Who used a different method of keeping track of the numbers
they used?*

Finally, ask the children if they can predict which numbers can
be halved or if they can see a pattern.

Can you see any pattern about which numbers can be halved?

What can you say about all even numbers? All odd numbers?

Assessing using and applying

- Children can halve an amount
  of counters and, as a result, say
  whether the number is odd or
  even.
- Children can identify odd and
  even numbers to at least 20.
- Children can make a
  generalisation about odd and
  even numbers and use this
  to say whether large numbers
  are odd or even.

Supporting the challenge

- Children work systematically
  through all the numbers to 10.
- Provide children with Numicon
  to see the structure of the
  numbers.

Extending the challenge

- Encourage the children to pose
  their own problems and extend
  the challenge above 30.
- Draw a horizontal line across
  the divided shape at the top of
  RS6, dividing it into four and
  explore which numbers can be
  quartered.

Order, order!

Using knowledge of place value to order numbers to 100

Using and applying

Reasoning

Describe patterns and relationships involving numbers; make predictions and test these with examples

Communicating

Present solutions to puzzles and problems in an organised way; explain decisions, methods and results in pictorial, spoken or written form, using mathematical language

Maths content

Counting and understanding number

- Explain what each digit in a two-digit number represents; partition two-digit numbers in different ways, including into multiples of 1 and 10
- Order two-digit numbers and position them on a number line

Key vocabulary

order, position, place value, smallest, largest, higher, lower, consecutive

Resources

- Large 0–20 number cards
- RS7 (for each child)
- Set of 0–100 number cards or RS8, one-minute timer or stopwatch (for each pair)

For *Supporting the challenge*:
- 100-grid

Can you put the number cards in order in one minute or less?

Introducing the challenge

 Shuffle the set of large 0–20 number cards and distribute the cards among 21 children in the class. Ask these 21 children to stand. The other children in the class tell those children where to stand in order from 0 to 20.

Who should start our order? Why? Where should Esther stand?

Who comes next? And next?

Lia has number 20. Where should she stand?

Once the numbers have been ordered, take the cards from the 21 children and ask them to sit back at their places.

Set one of the cards aside without showing anyone which number it is. Shuffle the remaining cards and make sure the children now know that they are not in order. Invite two children to the front to help you. Explain that the cards need putting back into order youare going do this by giving the cards that are less than 10 to the child to your right, and those 10 or greater to the child to your left. Turn over the card one at a time and ask the class to help you decide which child to give each card to.

Is this number more or less than 10?

Continue until all the numbers are divided into two sets.

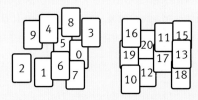

Next, ask the children to help you order the number cards from 0 to 20.

Which number comes first? Which one comes next?

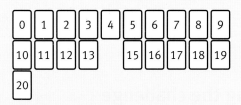

When the children have completed the order, they identify the card set aside: in this case, number 14.

The challenge

 Arrange the children into pairs and provide each child with a copy of RS7 and each pair with a set of 0–100 number cards and a one-minute timer or stopwatch. Children decide who is the Timekeeper. From the set of 0–100 number cards, each child sorts out a set of 11 consecutive numbers starting with a tens number: for example, 30 to 40. They shuffle these 11 cards, then swap their set for their partner's set. Can they put the cards in order in one minute? Children record the order of the cards on their copy of the resource sheet.

The challenge should develop fairly quickly into harder challenges. The children can sort out, shuffle and swap 11 consecutive cards not starting from a tens number: for example, 32 to 42.

Children can then shuffle the whole set of 0–100 cards and give each other 11 random cards. Each time, they see if they can put the cards in order in one minute or less. Each child checks their partner's sequence to see if it is in order. They record their sequences of cards on their resource sheet.

How are you going to order your set of numbers?

Which number is the smallest? Which one is the largest?

Drawing out using and applying

 Discuss with the children what strategies they used to decide which number was higher or lower, which numbers were consecutive, and how they could make judgements as quickly as possible.

Assessing using and applying

- Children can order a set of consecutive numbers starting with a tens number: for example, 30 to 40.
- Children can order a set of consecutive numbers not starting with a tens number: for example, 32 to 42.
- Children can order a random set of numbers within the range 0 to 100.

Supporting the challenge

- Provide the children with a 100-grid to help them put the numbers in order.
- Restrict the pack of number cards to numbers less than 20, 30, 40 or 50, depending on their ability.

Extending the challenge

- Children select out 0–9 digit cards from their pack and use these to make as many two-digit numbers as they can. They could pick two cards, record the number, repeat until all the cards have been used and then write the numbers in order.
- Children sort out a pack of 0–100 number cards according to various attributes, on which they decide themselves.

Pavements

Adding more than two numbers

Using and applying

Solving problems

Solve problems involving addition in the context of numbers

Representing

Identify and record the information or calculation needed to solve a puzzle or problem; carry out the steps or calculations and check the solution in the context of the problem

Maths content

Knowing and using number facts

- Derive and recall all addition facts for each number to at least 10

Calculating

- Add mentally a one-digit number to any two-digit number

Key vocabulary

addition, add, sum, total, plus, more, equals, altogether, smallest, largest, greatest

Resources

- NNS ITP: 20 cards
- Individual whiteboard and marker (for each pair)

For each child:

- RS9
- About 12 counters
- 1–6 or 0–9 dice

For *Supporting the challenge*:

- Number line

Which route from start to finish gives the greatest total?

Introducing the challenge

Arrange the children into pairs and provide each pair with an individual whiteboard and marker. Using the NNS ITP: 20 cards, turn on the grid and make a row of 10 cards with numbers from 1 to 6. Drag and drop two of the cards to the centre of the screen and reveal the two numbers.

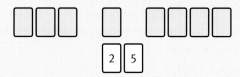

Working in pairs, children add them together and write the answer on their individual whiteboard.

What is 2 add 5?

How did you get that answer?

Did anyone work it out another way?

Who just knew straight away that the answer was ...?

Move the two cards back into the row at the top of the screen and repeat the above several times, asking children to add pairs of numbers. Occasionally, ask the children to explain how they worked out the answer.

Next, drag and drop three of the cards to the centre of the screen, reveal the three numbers and repeat the above.

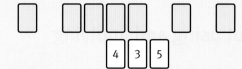

The challenge

Provide each child with a copy of RS9, about 12 counters and either a 1–6 or a 0–9 dice, depending on their ability. The children roll the dice repeatedly and write the number in any one of the 'paving stones' on the resource sheet. They continue rolling the dice until each space has a number in it.

The children mark out a pathway from 'Start' to 'Finish' by placing counters on spaces, stepping-stone style.

They then remove the counters one at a time to reveal the number underneath. They record the numbers and find the total.

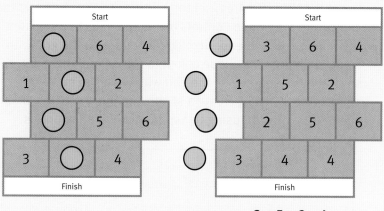

$$3 + 5 + 2 + 4$$

Challenge the children to find as many different routes as possible from 'Start' to 'Finish'. Which route gives the greatest total? Can they also find a route that gives the smallest total?

Drawing out using and applying

 Display a copy of RS9 on the interactive whiteboard or as an enlarged poster. Invite individual children to share the different routes they found with the class. As they read out their sequence of numbers, encourage the rest of the class to keep a running total of the numbers to check the answer.

Which numbers do you have on your paving stones?

Can you tell us one of your routes?

What is the total of your route?

What was the largest total you got? What numbers gave you that total? What was the smallest total?

Who found the greatest total? Which dice did you use? Why do you think you were able to make the greatest total?

Can you be sure that you investigated as many ways as possible?

What strategies did you use to try and make the total as large or as small as possible?

Assessing using and applying
- Children can add four or more numbers together.
- Children can find different routes to give them different totals.
- Children can find routes that give them the greatest and smallest totals.

Supporting the challenge
- Children use the 1–6 dice.
- Children use a number line to help them keep track of the calculation.

Extending the challenge
- Children use the 0–9 dice.
- Children create a different-shaped pavement to fill.

25p dog

Using known addition and multiplication facts

Using and applying

Representing

Identify and record the information or calculation needed to solve a puzzle or problem; carry out the steps or calculations and check the solution in the context of the problem

Communicating

Present solutions to puzzles and problems in an organised way; explain decisions, methods and results in pictorial, spoken or written form, using mathematical language and number sentences

Maths content

Knowing and using number facts

- Derive and recall all addition facts for each number to at least 10
- Derive and recall multiplication facts for the 2 times tables

Key vocabulary

addition, add, sum, plus, total, more, times, lots of, groups of, double, twice, altogether, equals, worth

Resources

- NNS ITP: Area or an OHP, an OHP copy of RS10 and transparent coloured counters.

For each pair:

- RS10, interlocking cubes in two colours
- Coloured pencil

For *Extending the challenge*:

- Interlocking cubes in a third colour

> Can you make a model dog that is worth exactly 25p?

Introducing the challenge

 Using the NNS ITP: Area, set the grid size to 5 by 5. Select the circle from the Shape Controls menu and select yellow from the Shape Colour menu. Highlight a square on the grid with a circle. Change the shape colour to purple and highlight several more squares on the grid with circles to make a shape. Alternatively, create a shape like the one below on an overhead projector, using transparent coloured counters.

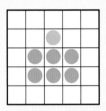

Explain to the children how the yellow circle stands for 2p and each of the purple circles stands for 1p. Ask them to calculate the value of the shape.

What is this shape worth?

How did you work it out? What did you count in?

Reset the ITP or rearrange your counters and repeat the above for other shapes. Encourage the children to talk about how they calculated the worth of each shape when there are more than two yellow circles in the shape. Did the children count in twos?

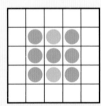 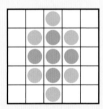

The challenge

 Arrange the children into pairs and provide each pair with a copy of RS10 and interlocking cubes in two colours. Each pair decides which colour is worth 1p and which is worth 2p. Children record their decision on the resource sheet.

Introduce the challenge to the class. Can pairs of children make a model dog using the cubes? Each dog must be worth exactly 25p. They need to find a way of recording their dog on the resource sheet and check that the total value of the cubes comes to 25p.

How are you going to start to make a 25p dog?

How else could you go about it?

How much is your dog worth?

Your dog is worth 27p, what could you change to make the dog worth exactly 25p?

How are you going to show this on your resource sheet?

Can you make another dog that is worth exactly 25p?

Drawing out using and applying

 Explore the different models the children have made. If appropriate, use the NNS ITP: Area to show some of these. As a class, check that each dog is worth 25p. Work with the children on efficient ways of checking: for example, by counting in twos or looking for combinations of cubes that total 10p.

Discuss the different combinations that the children have found. Are there others that they have missed?

Assessing using and applying

- Children can make a model dog that is worth 25p.
- Children can make a model dog that is worth 25p and record their dog accurately.
- Children can find different model dogs that are worth 25p.

Supporting the challenge

- Show the children how to draw a 2D representation of their dog.

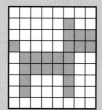

- Provide the children with 25p in coins. They use their money to 'buy' cubes so that they start with cubes that total 25p.

Extending the challenge

- Can the children make a family of dogs each worth 20p?
- Add in a third colour of cube for 3p or change the prices of the cubes.

Domino stars

Using known addition facts

Using and applying

Representing

Identify and record the information or calculation needed to solve a puzzle; carry out the steps or calculations and check the solution in the context of the problem

Communicating

Present solutions to puzzles in an organised way; explain decisions, methods and results in pictorial, spoken or written form, using mathematical language and number sentences

Maths content

Knowing and using number facts

- Derive and recall all addition facts for each number to at least 10
- Use knowledge of number facts and operations to estimate and check answers to calculations

Key vocabulary

addition, add, sum, plus, more, total, subtraction, subtract, take away, minus, difference between, less, altogether, equals, check

Resources

- Set of dominoes (for each pair)
- Supply of RS11 (for each child)

For *Supporting the challenge*:

- Number lines

> Can you find four dominoes that together have that number of dots?

Introducing the challenge

 Arrange the children into pairs and give each pair a set of dominoes. Children spread out their dominoes face up on the table in front of them. Explain that you are going to say a number from 2 to 12. The children have to find a domino with that number of dots and then tell you the two sets of dots on either side of the domino that gives that total. For example, you say the number 8 and the children find and say any of the following:

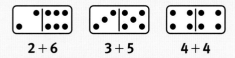

2 + 6   3 + 5   4 + 4

Repeat the above several times.

You may wish to briefly discuss with the children why more than one domino is often possible for giving the same total (as the above three examples show). Also discuss why each domino can represent two addition calculations:

2 + 6   6 + 2

Next, say a number between 0 and 6 and ask children to choose a domino that has this difference between the two sets of dots on either side. For example, you say the number 3 and the children find and say any of the following:

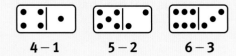

4 − 1   5 − 2   6 − 3

The challenge

 Ask a child to give you a number between 10 and 20 and write this on the board: for example, 15. Then ask the children working in pairs to find four dominoes that together have that total number of dots. Repeat asking the children to find a second set of four that has the same total.

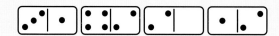

 Provide each pair with a supply of RS11 and give them a number between 15 and 20. Ask each child to find four dominoes that together have that number of dots. They put their dominoes on the resource sheet.

The children check that their partner's dominoes total the same number. They each record their set of dominoes and the total.

Discuss with the children the strategies that they used to check the total number of dots on their partner's dominoes. Encourage them to find quick methods of checking: for example, by looking for sets of dots that total 10.

Can you find four dominoes that total 17?

Your partner's four dominoes also total 17. What is different about their four dominoes?

How are you going to check that the total number of dots on your partner's dominoes is the same as yours?

Drawing out using and applying

 Share the different methods of recording the children have used, showing them how to use numerals and the addition sign if this was not done.

Who can tell me four dominoes that have a total of 18 dots?

Can anyone tell me another four dominoes that have a total of 18 dots?

Are there any more?

How did you show this on your resource sheet?

Did anyone show their dominoes in a different way?

Assessing using and applying

- Children can find four dominoes that add together to make a given total.
- Children can find different combinations of four dominoes to make a given total.
- Children can record different combinations of four dominoes that make a given total as number sentences.

Supporting the challenge

- Children use a number line to help them keep track of the calculations.
- Give the children a large number such as 18, 19 or 20 as this will give them a greater possibility of different combinations.

Extending the challenge

- Children choose a different target number and repeat the challenge several times.
- What is the largest possible total that a pair can find, using a standard dominoes set? What is the smallest total that they can find?

Line chunks

Using addition number facts and exploring number patterns

Using and applying

Reasoning

Describe patterns and relationships involving numbers; make predictions and test these with examples

Communicating

Present solutions to puzzles in an organised way; explain decisions, methods and results in pictorial, spoken or written form, using mathematical language and number sentences

Maths content

Counting and understanding number

- Describe and extend number sequences and recognise odd and even numbers

Knowing and using number facts

- Derive and recall all addition facts for each number to at least 10

Key vocabulary

addition, add, plus, sum, total, more, altogether, equals, pattern, sequence, odd, even, pair

Resources

- NNS ITP: Number line or one of the numbers tracks from RS13 (enlarged)

For each pair:

- Supply of RS12, 0–9 number tracks from RS13, blank number tracks from RS14

- Scissors, glue or sticky tape

Introducing the challenge

 Using the NNS ITP: Number line, decrease the maximum value to 10. Circle two consecutive numbers and ask the children for the total of these two numbers. Repeat several times occasionally asking children to explain how they worked out the answer.

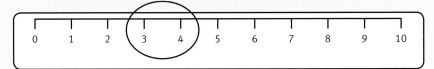

What is 3 add 4?

How did you work out that answer?

Did anyone work it out a different way?

If the children do not mention it, feed into the discussion the mental strategy of adding near doubles: for example, $6 + 7 = (6 + 6) + 1$ or $(7 + 7) - 1$

Did anyone use what they know about doubling to help them work out the answer? What did you do?

The challenge

 Arrange the children into pairs. Provide each pair with copies of RS12 and the other resources. Using one of the 0–9 number tracks from RS13, demonstrate that the number track can be cut up into chunks, each two pieces long. Explain to the children that they can either leave the zero on or cut it off: two different sets of chunks are then possible.

or

| 1 | 2 | | 3 | 4 | | 5 | 6 | | 7 | 8 |

Children cut the zero off a number track and cut it into two-piece chunks. They find the total of the two numbers, organise the chunks and stick them down on one copy of RS12 in a way that shows any patterns they have found, recording the totals next to the chunks.

Pairs repeat the challenge without first cutting the zero off the number track, using another copy of RS12.

Discuss the different totals, any patterns that children notice and the way they arranged their chunks.

Tell me one of the totals you made. Tell me the others.

What do you notice about these totals?

Can you see any pattern?

Why do you think this happened?

Drawing out using and applying

 Working in pairs, children use a blank number track and make part of a number line, starting with a number of their own choosing. When they have made the number track, they repeat cutting in chunks and adding as above, using a further copy of RS12.

Which number did you chose to start with?

What do you notice about your totals?

How do these compare with the other chunks you made?

Assessing using and applying

- Children can add pairs of consecutive numbers together and arrange them in a logical way.
- Children can see a pattern in the line chunks.
- Children can identify patterns and similarities in both types of line chunks and use these to make predictions about other pairs of numbers and their totals.

Supporting the challenge

- Assist the children in arranging their chunks on RS12 so that they can better identify patterns.
- Encourage children to model putting the larger number of each pair in their head and counting on the smaller number to reach the total.

Extending the challenge

- What happens if you cut the line into chunks three or four numbers long?
- Make different types of number tracks: for example,

2	4	6	8

and so on.

Money in my purse

Adding and subtracting twos and tens

Using and applying

Solving problems

Solve problems involving addition and subtraction in contexts of pounds and pence

Reasoning

Describe patterns and relationships involving numbers

Maths content

Counting and understanding number

- Describe and extend number sequences

Knowing and using number facts

- Derive and recall all addition and subtraction facts for each number to at least 10

Key vocabulary

addition, add, sum, total, plus, more, subtraction, subtract, take away, minus, difference between, less, equals, count, pattern

Resources

- NNS ITP: Number spinners

For each child:

- A purse containing 4p
- RS15

For each pair:

- 1–6 dice
- Spinner from RS16, paper clip and pencil for the spinner, supply of 2p and 10p coins (preferably real)

For *Supporting the challenge*:

- 1, 2, 3, 1, 2, 3 dice

What patterns do you notice?

Introducing the challenge

 Using the NNS ITP: Number spinners, make two six-sided spinners. Change the numbers on one of the spinners to 1 to 6 and the other to 2, 5, 2, 5, 2, 5. Spin both spinners. Pointing to each spinner in turn, tell the children that you want them to count on four steps of five from 0: that is, '0, '5' (show one finger), '10' (show two fingers), '15' (show three fingers), '20' (show four fingers). Demonstrate this to the children.

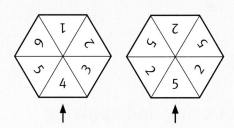

Spin the spinners and repeat again several times:

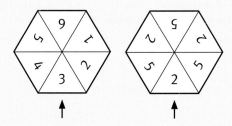

Three steps of two from 0:

'0', '2' (show one finger), '4' (show two fingers), '6' (show three fingers)

Six steps of two from 2:

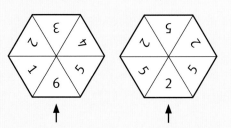

'0', '2' (show one finger), '4' (show two fingers), '6' (show three fingers), '8' (show four fingers), '10' (show five fingers), '12' (show six fingers)

The challenge

 Arrange the children into pairs and provide each child with a purse (or similar) containing 4p and RS15 and each pair with the other resources.

Children take turns to roll the 1–6 dice and spin the spinner. They collect that number of coins: for example, four 2p coins if they rolled 4 and spun 2p. Both children count together as one child drops the money into their purse: "2, 4, 6, 8." When both children have had a turn, they count all the money in their purses, including the 4p already there. They each record 4p in the first purse on their copy of RS15 and the total amount in the second.

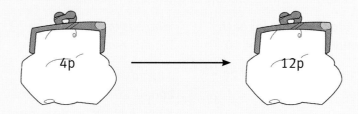

Children then put all the money back except the 4p, which goes back in the purse, and start again.

When the children have done this a few times, ask them to count on from the 4p already in their purses: "4, 6, 8, 10, 12" or "4, 14, 24, 34, 44" and record this straight away in the second purse.

What do you notice when you count on in twos from 4p?

What about when you count on in tens?

Drawing out using and applying

 Discuss with the children how they counted on in twos and tens. Place some money into one of the purses and ask the class to count with you the amount of money in the purse. Encourage the children to count in groups rather than in ones. Discuss the patterns of the numbers with them: "4, 14, 24 ..."

How did you keep track of your counting when you were counting on in twos?

What about when you were counting on in tens?

Assessing using and applying

- Children can count on in twos and tens.
- Children can identify patterns when counting on in twos and tens.
- Children can use their knowledge of counting in twos and tens to help them add and subtract twos and tens.

Supporting the challenge

- Provide the children with a 1, 2, 3, 1, 2, 3 dice.
- Make sure children use their purse and the real coins to actually put coins into the purse before using the resource sheet.

Extending the challenge

- Use a different starting amount in the purse.
- Children have two turns each, adding their second score to their first.

Making 20

Writing calculations for a given total

Using and applying

Solving problems

Solve problems involving addition and multiplication in contexts of numbers

Representing

Identify and record the calculation needed to solve a puzzle; carry out the steps or calculations and check the solution in the context of the problem

Maths content

Knowing and using number facts

- Derive and recall all addition facts for each number to at least 10, all pairs with totals to 20 and all pairs of multiples of 10 with totals to 100
- Recall doubles of all numbers to 20
- Derive and recall multiplication facts for the 2 times tables

Key vocabulary

addition, add, sum, total, plus, more, multiplication, multiply, times, lots of, groups of, altogether, equals

Resources

- Set of 0–19 number cards
- Set of multiples of 10 number cards from 10 to 100
- Additional 50 number card
- RS17 (for each child)

For *Supporting the challenge*:

- Number lines

How many different ways can you score 20 points, using three darts?

Introducing the challenge

 Distribute the 31 number cards among the class. If there are more cards than children, give some children two cards. Explain that you are going to revise their knowledge of addition, doubling and multiplication facts. Tell them that you are going to say a calculation and the child who has that answer on their card holds up the number card.

Ask questions similar to the following that consolidate children's understanding of the following facts:

Addition facts for each number to at least 10	*Who has the sum of 4 and 3? What is 6 add 2?*
Pairs with totals of 20	*12 and what other number equals 20? 8 and how many more makes 20?*
Pairs of multiples of 10 with totals of 100	*40 and what equals 100? 20 plus what other number totals 100?*
Doubles of all numbers to 20.	*What is double 6?*
Multiplication facts for the 2 times table	*What is 4 times 2?*

The challenge

 Give each child a copy of RS17 and display the resource sheet on the interactive whiteboard or as an enlarged poster. Introduce the challenge to the children. The investigation is self-explanatory, but make sure that the children realise how the rule works. For example, scores of 20 can be achieved as follows:

$$8 + 7 + 5 \quad (2 \times 4) + 7 + 5$$

$$9 + 9 + 2 \quad (2 \times 8) + 1 + 3$$

It is not necessary at this stage for children to use brackets in a calculation. However, if you feel this is appropriate for some children, you may wish to introduce them to this convention.

 Give the children enough time to work on the investigation independently.

Tell me one way of making 20.

Can you tell me a way that uses the number 9?

What about using a double?

How are you going to record this as a number sentence?

 Arrange the children into pairs and ask them to compare their results.

Drawing out using and applying

 Bring the class back together again and invite individual children to share some of their calculations and one of their partner's calculations they thought was of interest with the rest of the class.

Tell us one of your number sentences that involves doubling/ multiplying by 2.

Who found 10 different number sentences?

Who found more than 10? More than 20?

Draw the discussion round to the way in which children recorded their results. If any children used brackets, ask them to explain some of these to the rest of the class.

How did you write this out? Did you write a number sentence?

Tell us a number sentence involving doubling/multiplying by 2.

Who wrote out a similar number sentence in a different way?

Did anyone use any other methods of recording their number sentences?

Assessing using and applying
- Children can find different scores for 20 by adding three numbers.
- Children can use addition and doubling or multiplying by 2 to find different scores for 20.
- Children can find numerous ways of making 20, using mathematical symbols to record number sentences, including brackets.

Supporting the challenge
- Provide children with a number line to help with their calculations.
- Make the score 10 instead of 20.

Extending the challenge
- What other points can the children score, using three darts?
- What if the children used four darts?

Paul's pets

Solving a problem involving addition

Using and applying

Solving problems

Solve problems involving addition in contexts of numbers

Representing

Identify and record the information or calculation needed to solve a puzzle; carry out the steps or calculations and check the solution in the context of the problem

Maths content

Knowing and using number facts

- Derive and recall all addition facts for each number to at least 10
- Use knowledge of number facts and operations to estimate and check answers to calculations

Key vocabulary

addition, add, sum, total, plus, altogether

Resources

For each child:
- RS18
- 12 interlocking cubes

For *Supporting the challenge*:
- Additional interlocking cubes

What pets might Paul have?

Introducing the challenge

 Introduce the challenge to the children by holding a discussion on the similarities and differences between animals: for example, how they move, their outer covering, where they live, what they eat, and so on. Lead the discussion towards the number of legs they have. Ask the children for examples of animals with two, four, six and eight legs. Write some of these on the board.

2 legs	4 legs	6 legs	8 legs
birds	cats dogs mice	insects	spiders

Ask the children questions similar to the following, regarding the total number of legs there are in different groups of animals:

How many legs do three birds have?

How many legs do two birds and one insect have?

How many legs are there altogether if there are two cats, three insects and one spider?

The challenge

 Provide each child with a copy of RS18 and 12 or more interlocking cubes. Read through the challenge with the children. Invite children to express the challenge in their own words.

Children collect 12 interlocking cubes together to represent 12 legs. Discuss the different number of combinations of pets that Paul might have: that is, two pets, three pets, four pets or more. Make sure that the children are happy that a collection of pets could contain all three different types of animals or just two or possibly the same pet over and over again.

Ask the class for a possible solution and illustrate this, using the interlocking cubes.

Does this combination use all 12 legs?

Explain to the children that they need to think carefully about how they are going to organise and present their findings in order to find all the different possibilities.

 Give the children enough time to work on the challenge independently. Monitor the children as they work.

How are you going to record this combination?

Look at this combination of pets. How could you change it slightly to come up with another possibility?

What patterns do you notice?

Drawing out using and applying

 Invite individual children to say one or two possible combinations. Continue until all seven combinations have been identified. (Mathematically, there are seven combinations, but the children of course might think there are more than this: for example, an insect, mouse and bird is different from an insect, dog and bird.) Discuss duplications if these arise: for example, insect, mouse, bird (6 + 4 + 2) and mouse, bird, insect (4 + 2 + 6).

What is one collection of pets that Paul might have?

What is another?

What other ones include birds?

How are these combinations the same?

Lead the discussion to how children kept a record of their results. Invite individual children to show and explain their methods. Discuss the benefits and limitations of these.

How did you keep a record of what you did?

Who used a different method?

Finally, ask the children if they noticed any patterns as they were working which helped them identify all the different possible combinations: for example, starting with the pet with the greatest or smallest number of legs.

Did anyone use a system as they were working to make sure that they found all the different combinations? What did you do?

Did anyone have a different system?

Assessing using and applying

- Children can find different combinations of pets.
- Children can find different combinations of pets and record these on paper.
- Children can work logically and record their results systematically to find all the different possible combinations of pets.

Supporting the challenge

- Provide the children with extra interlocking cubes and allow them to use these to show all the different combinations they can think of. This will help them identify any duplication and also, if appropriate, to later record their results.
- Help children keep a record of how many different pets Paul might have: for example, 2 pets, 3 pets, 4 pets, 5 pets, 6 pets.

Extending the challenge

- Do not provide the children with interlocking cubes and ask them to solve the problem, using their own method.
- Children make a similar problem for a friend to solve, using the list of pets written on the board.

Number bracelets

Recognising addition and subtraction as inverse operations

Using and applying

Solving problems

Solve problems involving addition and subtraction in numerical contexts

Representing

Identify and record the information or calculation needed to solve a problem; carry out the steps or calculations and check the solution in the context of the problem

Maths content

Calculating

- Add or subtract mentally a one-digit number to or from another one-digit number
- Understand that subtraction is the inverse of addition and vice versa and use this to derive and record related addition and subtraction number sentences

Key vocabulary

addition, add, sum, plus, total, altogether, subtraction, subtract, minus, take away, difference between, more, less, equals

Resources

- 11 individual whiteboards
- 2 whiteboard markers in different colours

For each pair:
- RS19
- Calculator

For *Supporting the challenge*:
- Cubes or counters

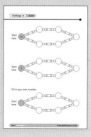

> **How did you complete the bracelet so that you ended up back at the starting number?**

Introducing the challenge

 Ask 11 children to come to the front of the class and stand in a line. Give each child an individual whiteboard. Using one of the whiteboard markers, write a number less than 10 on the first whiteboard in the line, such as 8, a number below 10 on the third whiteboard, such as 3, and the same number as is written on the first whiteboard on the last whiteboard in the line:

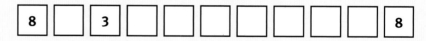

Children suggest how to get from 8 to 3. Take suggestions from the class before using the other whiteboard marker to write the operation and number on the second whiteboard:

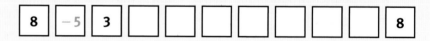

Write another number on the fifth whiteboard and continue in the same way until you reach the last whiteboard:

Rub out all the numbers and operations from the whiteboards and repeat, inviting a different group of 11 children to take a whiteboard each.

The challenge

 Provide each pair with RS19 and a calculator. Display a copy of the resource sheet on the interactive whiteboard or as an enlarged poster. Children enter the number 10 into their calculator. Ask them whether to add or subtract and press this key (subtract in the example). Ask for a number below 10 (4 in the example) and enter this into the calculator. Write the result of the calculation on the resource sheet in the next circle on the first bracelet.

*Checking results in a different way,
predicting and justifying*

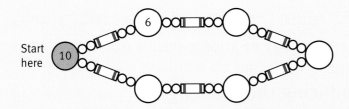

Discuss the result. You may choose to write the operation over
the bracelet connecting the two circles.

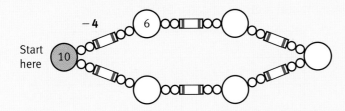

Ask for another operation and another number to complete
a third number on the bracelet.

Tell the children that to complete the number bracelet
successfully, they need to end up back at the starting number.
Ask for ideas for returning to 10 in the last two 'jumps'. Try out
the children's ideas.

What operation should we do so that we can get close to 10?

What else could we do?

What do we need to do now so that we reach 10?

 Encourage the children to find their own ways of completing
the two remaining bracelets on their resource sheet.

Drawing out using and applying

 Arrange the pairs into groups of fours. Children exchange
their results and check each other's calculations.

 Look at some different bracelets. Encourage the children
to discuss them and explain to the rest of the class how they
set about solving the challenge and justifying their answers.

Which bracelet do you think is the most interesting? Why?

Can you see any operations 'undoing' each other?

Assessing using and applying

- Children can return to the starting number after several calculations.
- Children can demonstrate and explain different ways of 'undoing' calculations.
- Children can recognise, explain and use inverse operations.

Supporting the challenge

- Children use cubes or counters to help with their calculations.
- Tell the children to use numbers to 5.

Extending the challenge

- Children choose the same starting number and come up with at least three different bracelets from the same starting point.
- Children explore what would happen with longer or shorter bracelets.

Collect the coins

Adding more than two numbers and making estimates

Using and applying

Reasoning

Describe patterns and relationships involving numbers; make predictions and test these with examples

Communicating

Present solutions to puzzles in an organised way; explain decisions, methods and results in pictorial, spoken or written form, using mathematical language and number sentences

Maths content

Knowing and using number facts

- Derive and recall all addition facts for each number to at least 10

Calculating

- Add mentally a one-digit number to any two-digit number

Key vocabulary

addition, add, sum, total, plus, more, altogether, equals

Resources

- RS20

For each child:

- RS21, 1–6 dice, cube

For *Supporting the challenge*:

- Counters
- Number lines

For *Extending the challenge*:

- 0–9 dice

Find a route through the castle that allows you to collect as many coins as possible.

Introducing the challenge

 Display RS20 on the interactive whiteboard or as an enlarged poster. Circle a vertical, horizontal or diagonal line of any three numbers and announce the total to the children, explaining your mental method of calculation.

5	4	6	3	2	1	5	1	6	3
1	6	2	4	5	3	2	5	1	4
5	3	1	3	5	2	4	2	4	1
2	5	1	3	4	5	5	3	1	9
3	0	4	2	6	1	5	2	4	6
4	1	5	6	4	1	2	3	3	5
0	1	3	4	3	5	1	2	6	6
5	2	3	1	4	2	6	4	5	3
2	1	0	5	4	4	3	6	1	2
5	3	5	4	1	2	0	3	4	2

Five and five makes 10, and four more is 14.

Invite a child to come to the board and circle three numbers, say the total and explain how they worked it out. Repeat several times before asking children to circle four, then five numbers.

Can anyone find the total of six or more numbers?

The challenge

 Provide each child with a copy of RS21 and the other resources. Display RS21 on the interactive whiteboard or as an enlarged poster. Ask the children to imagine that they are magicians who are going to place some gold coins in each room of the castle. Children place their cube in a room and roll the dice to decide how many coins to place in that room. They remove the cubes and record the number of coins placed in that room on the resource sheet. They repeat this until all nine rooms have a number of coins in them.

Introduce the challenge to the children. Explain that they are going to travel through the castle, collecting coins as they go. The magician has cast a spell that means each room can be visited only once: going back into a room has dire consequences (the children can decide what!). They can collect all the coins in a room during one visit.

Can the children find a way through the castle, going in and out at the marked points, visiting any room once only and collecting as many coins as possible as they go? Encourage them to record their totals and routes in the scroll at the bottom of the resource sheet.

Describe your route through the castle.

How many coins did you collect in each room?

What was the total number of coins you collected?

Do you think you can find another route that will give you even more gold coins?

Drawing out using and applying

 Discuss the various totals that the children were able to collect. Individuals talk through the route they followed. As they read out the number of coins in each room visited, encourage the other children to keep a running total to check the final amount.

What was the greatest number of coins you collected?

Did anyone collect more gold coins than this? Describe your route to us.

Tell us how many coins you collected in each room.

What strategies did children use to maximise the number of coins they collected?

Did anyone find a way of visiting all the rooms?

Assessing using and applying

- Children can find a route through the castle and calculate the total number of coins collected.
- Children can find different routes through the castle and accurately calculate the total number of coins collected on each route.
- Children can find a route through the castle visiting all the rooms.

Supporting the challenge

- Give children a pile of counters. They put counters in each room corresponding to the number rolled on the dice.
- Provide children with a number line to help with adding the numbers.

Extending the challenge

- Children roll the dice and decide which room to place that number of coins in.
- Provide the children with a 0–9 dice instead of a 1–6 dice.

Calculating digit cards

Add and subtract a one-digit number to or from any two-digit number

Using and applying

Representing

Identify and record the calculation needed to solve a puzzle; carry out the steps or calculations and check the solution in the context of the problem

Communicating

Present solutions to puzzles in an organised way; explain decisions, methods and results in pictorial, spoken or written form, using mathematical language and number sentences

Maths content

Calculating

- Add or subtract mentally a one-digit number to or from any two-digit number
- Use the symbols +, − and = to record and interpret number sentences

Key vocabulary

digit, number sentence, calculation, addition, add, sum, total, plus, subtraction, subtract, minus, take away, leaves, more, less, altogether, equals

Resources

- Set of 1–9 demonstration cards or NNS ITP: 20 cards

For each pair:
- RS22
- Set of 1–9 digit cards

How many number sentences can you make, using the three cards?

Introducing the challenge

 If using demonstration cards, shuffle the cards, turn over in the top three and use these to make a two-digit and a one-digit number as described below. If using the NNS ITP: 20 cards, make a stack of 9 cards, with 1 as the first card number, 1 as the step number and 0 as the increment number. Turn on the grid and deal the cards. Choose three of the cards and drag and drop them to the centre of the screen arranged as a two-digit number and a one-digit number.

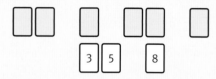

Ask the children for the sum of the two numbers and write this as a number sentence on the board:

$$35 + 8 = 43$$

Briefly ask individual children how they worked out the answer.

How did you get that answer?

Did anyone work it out using a different method?

If appropriate, work out the answer with the children, using, for example, an empty number line, and write the calculation underneath.

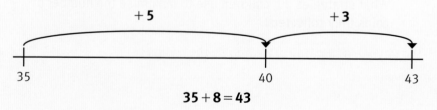

$$35 + 8 = 43$$

The challenge

Arrange the children into pairs and provide each pair with a copy of RS22 and a set of 1–9 digit cards. Introduce the challenge to the children, drawing parallels between the challenge and the activity they have just done on the interactive whiteboard. Encourage the children to think carefully about how they are going to record their different calculations and to try and look out for a pattern or system that will help them identify all the different calculations that are possible. There are six different addition calculations and six different subtraction calculations possible.

What is one addition number sentence you can make using these three digits? Can you tell me another one?

Can you see a pattern that will help you make sure that you find all the different number sentences possible?

Drawing out using and applying

 Bring the class back together and using either the demonstration cards or the NNS ITP: 20 cards, ask one pair of children to tell you the three numbers they chose. Display the demonstration cards or drag and drop these cards to the centre of the screen. With the help of the class, write down all the different addition and subtraction number sentences possible. As you do this, encourage the children to look for patterns and a system which helps them identify all the different calculations possible.

$$\boxed{3}\ \boxed{5}\qquad \boxed{2}$$

$53 + 2 = 55$	$53 - 2 = 51$
$52 + 3 = 55$	$52 - 3 = 49$
$35 + 2 = 37$	$35 - 2 = 33$
$32 + 5 = 37$	$32 - 5 = 27$
$25 + 3 = 28$	$25 - 3 = 22$
$23 + 5 = 28$	$23 - 5 = 18$

How many different addition number sentences are there? What about subtraction number sentences?

Can you see a system? Look at the column showing the first number in each number sentence. What do you notice?

Finally, referring back to the calculations on the whiteboard, ask the children to order the answers from smallest to largest.

Assessing using and applying

- Children can use the cards to make addition and subtraction number sentences.
- Children can arrange the cards to make six different addition calculations and order the answers from smallest to largest.
- Children can work systematically and arrange the cards to make six different addition and six different subtraction calculations and order the answers from smallest to largest.

Supporting the challenge

- Draw children's attention to the pattern in the addition calculations: that is, answers appearing twice.
- Make them aware of the strategy of arranging the digit cards to make the largest possible two-digit number first, then working down to the smallest two-digit number.

Extending the challenge

- Children choose three different digit cards and investigate what other two-digit plus one-digit calculations they can make, using three of the cards at a time.
- What if the children chose four cards and arranged them to make two two-digit numbers? How many different addition and subtraction calculations could they write?

Number families

Identifying and recording related addition and subtraction number sentences

Using and applying

Representing

Identify and record the information or calculation needed to solve a problem; carry out the steps or calculations and check the solution in the context of the problem

Reasoning

Describe patterns and relationships involving numbers; make predictions and test these with examples

Maths content

Knowing and using number facts

- Derive and recall all addition and subtraction facts for each number to at least 10

Calculating

- Understand that subtraction is the inverse of addition and vice versa and use this to derive and record related addition and subtraction number sentences
- Use symbols to record and interpret number sentences

Key vocabulary

digit, number sentence, calculation, addition, add, sum, total, plus, subtraction, subtract, minus, take away, leaves, more, less, altogether, equals

Resources

- Large 1–10 number cards
- Supply of RS23 (for each child)

For *Supporting the challenge*:

- 10 interlocking cubes

What addition and subtraction number sentences can you make with three numbers?

Introducing the challenge

 Give the number cards to 10 children in the class. Say two numbers: for example, 4 and 6. The two children holding these cards stand up, holding their cards for all the class to see. Say either an addition or subtraction statement where the answer will be 10 or less.

Add these two numbers together. What is the sum/total of these two numbers?

How much more is 6 than 4? What is the difference between these two numbers?

The child holding the number card that is the answer to your statement calls out the number and stands up, holding the card for the class to see.

Repeat several times with the same set of children before asking these children to pass their number card to a child that has not previously had a card. Repeat several times. If necessary, repeat again for a third set of children.

The challenge

 Draw a large triangle on the board and write the following numbers in each of the three corners:

Explain to the children how these numbers can be used to make two addition and two subtraction number sentences.

$$3 + 2 = 5$$
$$2 + 3 = 5$$
$$5 - 2 = 3$$
$$5 - 3 = 2$$

Three add two equals five. Who can tell me another addition number sentence that uses the same three numbers?

Who can tell me a subtraction number sentence?

Can we make any more addition and subtraction number sentences using these three numbers?

If appropriate, repeat the above for another set of three numbers.

 Provide each child with a supply of RS23. Children record other sets of three numbers that can be used to make two addition and two subtraction number sentences.

Can you tell me three numbers that go together to make a number family?

What are the addition number sentences? What are the subtraction number sentences?

Five, four and what other number completes this number family? What number sentences can you make using these three numbers?

Drawing out using and applying

 Provide an opportunity for children to report back what they found out. Display an example of three numbers that do not make two addition and two subtraction number sentences.

5   3   9

What is wrong with this number family?

Which number could I change to make this a proper number family? Could I change a different number?

Discuss with the children the inverse relationship between addition and subtraction, asking them to think of times when it might be useful to know about this relationship.

Assessing using and applying

- Children can list three numbers that make two addition and two subtraction calculations.
- Children can see that addition 'undoes' subtraction and vice versa.
- Children understand the inverse relationship between addition and subtraction and can use this to derive and record a range of addition and subtraction calculations.

Supporting the challenge

- Start the children off by providing them with two of the numbers in a number family and asking them to identify the third number.
- Children take 10 or fewer interlocking cubes and make them into a rod. They then split the rod in two.

$3 + 4 = 7$

$4 + 3 = 7$

$7 - 3 = 4$

$7 - 4 = 3$

Extending the challenge

- Children choose a number greater than 10 and find several solution sets that each include their chosen number
- Can the children see a similar relationship between multiplication and division and provide examples to illustrate this?

Equal groups

Understanding division

Using and applying

Representing

Identify and record the information or calculation needed to solve a problem; carry out the steps or calculations and check the solution in the context of the problem

Reasoning

Describe patterns and relationships involving numbers; make predictions and test these with examples

Maths content

Calculating

- Represent repeated subtraction (grouping) as division; use practical and informal written methods and related vocabulary to support division
- Use the symbols ÷ and = to record and interpret number sentences

Key vocabulary

multiplication, multiply, times, lots of, groups of, division, group

Resources

- NNS ITP: Number dials
- 2 rulers

For each pair:
- RS24 (optional)
- About 40 interlocking cubes

For *Supporting the challenge:*
- RS25

> # How many different ways can you find to put 36 cubes into equal groups?

Introducing the challenge

 Introduce 'Multiplication challenge' to the children. Using the NNS ITP: Number dials, set the number in the middle to 2 and show the number. Hide the numbers around the edge of the dial and show all the products.

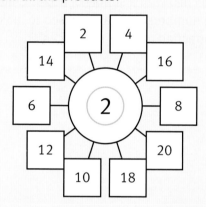

Invite a child to come to the front and try their multiplication challenge. Provide the volunteer with the ruler, as commodification calculation of the 2 times table: for example, "What is 2 × 8?" The child points to the answer with the ruler. Can they get them all right without making any mistakes? If appropriate, challenge the children, using other times tables facts such as the 5 or 10 times tables.

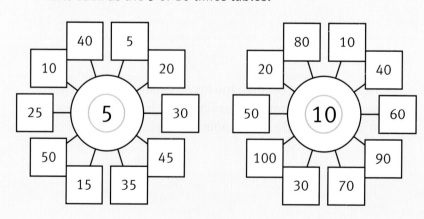

The challenge

 Arrange the children into pairs and provide each pair with a copy of RS24 and about 40 interlocking cubes. Introduce the challenge to the children. Children count out 36 cubes and recount the cubes to check that they have exactly 36. They put aside any additional cubes.

 Working in pairs, children find as many different ways as they can to put 36 cubes into equal groups, with none left over. As they work through the challenge, encourage them to think about the best way to record their results.

How are you going to keep a record of that combination?

How many groups have you made? How many cubes are there in each group?

How could you record this as a number sentence?

What other number sentence could you write that uses the same three numbers? What does this mean? How are these two calculations different?

Are there any other ways you could put the 36 cubes into equal groups?

Drawing out using and applying

 Bring the class back together again and ask individual pairs to say one or two of the combinations they found. Discuss with the children the different ways they kept a record of the combinations they found. When the opportunity arises, discuss the similarities between pairs of calculations: for example, $36 \div 9 = 4$ and $36 \div 4 = 9$.

How many different ways did you find to put 36 into equal groups?

If you know that you can put 36 cubes into three groups of 12, how many groups would you have if you put the 36 cubes into groups of three?

How did you record this combination?

Did anyone record their results in a different way?

Who wrote out their results as a number sentence? Explain to us what they mean.

Assessing using and applying

- Children can put 36 cubes into different groups of equal sizes.
- Children can find all the different ways of putting 36 cubes into equal groups.
- Children can write the corresponding division calculation.

Supporting the challenge

- Using RS25, children circle 36 dots into equal groups. How many different ways can they do this?

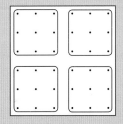

- Children find different ways to put 12 dots into equal groups.

Extending the challenge

- What if you used 24 or 48 cubes?
- Which numbers of cubes can you not divide into equal groups?

Remainder 1

Understanding the concept of remainders

Using and applying

Representing

Identify and record the calculation needed to solve a puzzle; carry out the steps or calculations and check the solution in the context of the problem

Communicating

Present solutions to puzzles in an organised way; explain decisions, methods and results in pictorial, spoken or written form, using mathematical language and number sentences

Maths content

Calculating

- Use practical and informal written methods and related vocabulary to support division, including calculations with remainders
- Use the symbols ÷ and = to record and interpret number sentences

Key vocabulary

number, one-digit, two-digit, division, divide, share, group, remainder

Resources

For each pair:

- RS26 (optional)
- Pile of counters

For *Supporting the challenge*:

- Yoghurt pots or similar

What other ways can you group a set of counters so that there is a remainder of 1?

Introducing the challenge

 Invite nine children to come and stand in a row at the front of the class. Ask the first two children to hold hands, then ask the next two children to hold hands, and so on, until there are four pairs of children holding hands and one child by themselves.

Discuss with the class how the nine children have been arranged. Establish that they have been put into groups of two and that there are four groups of two children and one child left over (remaining). At this stage, do not record this as a number sentence.

Repeat the above for other groups of children, arranging them into equal groups with one or more children left over: for example, nine children arranged into groups of four, 11 children arranged into groups of five and 13 children arranged into groups of three.

The challenge

 Arrange the children into pairs. Provide each pair with a copy of RS26 if appropriate and a pile of counters. Introduce the challenge to the children, using the example of the 19 counters. Explain that they can use any number of counters and arrange them into groups of any size, but they have to make sure that there is one counter left over each time.

Emphasise to the children that part of this challenge is for them to decide upon the best way to keep a record of their work. If appropriate, discuss with the children the use of the symbols ÷ and = to record number sentences. Alternatively, you may wish to leave this for the children to suggest themselves as they work through the challenge.

 Monitor the children as they work on the challenge. If pairs of children seem to be experiencing difficulty in finding different ways of grouping the counters so that there is a remainder of 1, remind them of their knowledge of times tables facts and those numbers that would produce a remainder of 1: for example, $3 \times 10 = 30$, so $31 \div 10 = 3$ remainder 1

What other ways can you group a set of counters so that there is a remainder of 1?

How are you going to keep a record of this?

This gives two left over. What can you change so that there will only be one left over?

If you know that six counters put into groups of two gives three groups, how many counters would you need so that there were three groups of two counters and one left over?

Can you write this as a number sentence?

Drawing out using and applying

 Bring the class back together and ask individual pairs to suggest some of the groupings they made. Also ask them to explain to the rest of the class how they recorded this. Where you have observed pairs of children using the symbols ÷ and = to record number sentences, invite them to share this with the rest of the class. Ask the class if anyone used their knowledge of their times tables facts to help them find ways of grouping a set of counters so that there is a remainder of 1.

Tell us one of the ways you found of grouping a set of counters so that there was a remainder of 1.

How did you record this?

Did anyone record their groupings as a number sentence? Explain to us what you wrote.

How did knowing some of your times tables help you think of ways of grouping counters so that there was a remainder of 1?

Assessing using and applying

- Children can find ways of grouping a set of counters so that there is a remainder of 1.
- Children can find ways of grouping a set of counters so that there is a remainder of 1 and record this as a number sentence.
- Children can use their knowledge of known multiplication and division facts to work out calculations with remainders of 1.

Supporting the challenge

- Provide the children with 12 or fewer counters.
- Provide the children with yoghurt pots or similar to help them arrange the counters into different groups.

Extending the challenge

- Encourage the children to record their groupings as number sentences.
- What about calculations with a remainder of 2 or 3?

Name snakes

Solving problems involving multiplication and division, including those with remainders

Using and applying

Reasoning

Describe patterns and relationships involving numbers; make predictions and test these with examples

Communicating

Present solutions to puzzles and problems in an organised way; explain decisions, methods and results in pictorial, spoken or written form, using mathematical language and number sentences

Maths content

Knowing and using number facts

- Derive and recall multiplication and the related division facts; recognise multiples

Calculating

- Use practical and informal written methods and related vocabulary to support multiplication and division, including calculations with remainders

Key vocabulary

multiplication, multiply, times, lot of, groups of, division, divide, share, group, equally, equals, remainder, left over

Resources

- NNS ITP: Area

For each child:

- RS27
- Supply of RS28
- Scissors

> **Does your name fit into the snake an exact number of times? How many are left?**

Introducing the challenge

 Using the NNS ITP: Area, set the grid size to 10 by 10. Select the circle from the Shape Controls menu and select yellow from the Shape Colour menu. Hide the grid line or pinboard. Highlight an even number of circles on the (hidden) grid. Alternatively, place a blank transparency on an overhead projector and put eight counters on top of it. Continue as below.

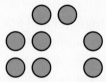

Tell the children that you are going to place these circles into groups of two. Circle the circles into groups of two. Explain to the children that there are eight circles and that you have put them into four groups of two.

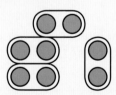

Remove the four rings and tell the children that this time you want to put them into groups of three. Invite a child to the board to do this and discuss the results. Explain that there are eight circles and that these are two groups of three, with two left over.

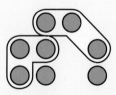

Repeat the above several times with 10 or 12 circles, either by resetting the ITP or changing the number of counters on the overhead projector.

The challenge

 Give each child a copy of RS27. Children count how many spaces long the snake is. Explain that they are going to write their name repeatedly, one letter per space, until the snake is filled. Before they start, ask if anyone thinks that their name will fit into the snake an exact number of times.

Who thinks their name will fit into the snake an exact number of times?

Why do you think that?

Who thinks that they won't be able to finish writing their name by the time they get to the end of the snake? What makes you think that?

 Allow enough time for the children to complete the first part of the challenge.

 Once the children have filled the snake, find out and discuss whether anyone's name filled it exactly. Can anyone change or adapt their name so that it would fit exactly: for example, Mike for Michael? Where someone's name did not fit exactly, discuss the number of spaces left over or the number of letters in the last incomplete name.

 Provide each child with a supply of blank snakes from RS28 and scissors and challenge them to cut the snakes or shade in spaces from the tail, to form lengths so that their names will fit in an exact number of times. Encourage the children to work out and predict how long the snakes need to be before creating them, rather than simply filling in their names and then shading or cutting off spare spaces. Can the children make more than one name snake, each a different length?

Drawing out using and applying

 Discuss the relationship between the number of spaces in the snakes and the number of letters in the names.

Roxanne, which other numbers of spaces will work for your name?

Are there lengths of snakes that work for several names? Why is this?

Are there lengths of snakes that do not work for anybody?

Assessing using and applying

- Children can make snakes of different lengths where their name fits in an exact number of times.
- Children can see a link between the number of spaces in the snakes and the number of letters in the names.
- Children can use the relationship between the number of spaces in the snakes and the number of letters in the names to predict which other numbers of spaces will work for their name.

Supporting the challenge

- Children work in pairs, checking each other's name snakes as they make them.
- Arrange children with the same number of letters in their names into pairs (for example, David and Lemar) and ask them to work together. What do they notice?

Extending the challenge

- Can the children write multiplication and division number sentences relating to their name snakes? What about the first snake they made from RS27? What do they do with the number of letters in the last incomplete name?
- Children turn their snakes into number tracks by numbering the spaces from 1 onward. They repeatedly write in their names as before and shade each space that contains their first-name initial. What patterns of numbers does this generate? What will be the next number in the sequence? Can they continue the pattern?

Calculating the value of an unknown in a number sentence

Using and applying

Representing

Identify and record the information or calculation needed to solve a problem; carry out the steps or calculations and check the solution in the context of the problem

Communicating

Present solutions to problems in an organised way; explain decisions, methods and results in pictorial, spoken or written form, using mathematical language and number sentences

Maths content

Knowing and using number facts

- Use knowledge of number facts and operations to estimate and check answers to calculations

Calculating

- Use the symbols $+$, $-$, $\times$, $\div$ and $=$ to record and interpret number sentences
- Understand that subtraction is the inverse of addition and vice versa

Key vocabulary

number, addition, subtraction, multiplication, division, equals, number sentence, symbol, sign

Resources

- 0–10 number cards, multiples of 2, 5, 10 number cards, maths symbol cards
- RS29 (for each child)
- Scissors (for each pair)
- Number lines

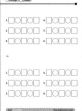

How did you work out the missing numbers and symbols?

Introducing the challenge

Invite five children to the front of the class and ask them to stand in a row, facing the class. Give the children a set of cards in the order so that they can make a number sentence. However, tell them not to show the rest of the class what is on their cards.

Tell the first four children in the row to reveal their cards to the class. The rest of the class tries to work out what is on the other card.

What do you think is on Derek's card?

What makes you say that?

Ask the child to reveal their card.

Repeat the above several times, using other combinations of cards to make different addition, subtraction, multiplication and division number sentences, including those that require the children to calculate the value of an unknown in a number sentence.

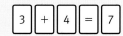

How did you work out that missing number/sign?

The challenge

Arrange the children into pairs and provide each child with a copy of RS29 and each pair with scissors. Working individually at first, children write down six different number sentences at the top of their resource sheet, a number in each square and a symbol in each circle. You may wish to direct the children to use only addition and subtraction or all four operations. If necessary, give children some indication as to which number facts to use: for example, addition and subtraction facts for each number to at least 10 and multiplication facts for the 2, 5 and 10 times tables and the related division facts. Children keep their six number sentences secret from their partner.

When they have written their six number sentences, children rewrite their facts at the bottom of their resource sheet, but this time omitting either one of the numbers or symbols from each of their number sentences.

Children then cut along the line of their resource sheet. They retain the top section of their sheet and give the bottom section to their partner. They work out the missing numbers and symbols from their partner's six number sentences. Once they have finished, they work together, comparing their results with the top sections of both resource sheets.

Drawing out using and applying

 Pairs of children share some of their calculations with the rest of the class. Move on the discussion to how children worked out the value of the unknown number or symbol and also discuss the concept of inverse relationships, how subtraction is the inverse of addition (and vice versa) and how children can use this to derive and record related addition and subtraction number sentences. If appropriate, also explain to the children the inverse relationship between multiplication and division, using known multiplication facts for the 2, 5 and 10 times tables and the related division facts as an example.

How did you work out what was missing in this number sentence?

Assessing using and applying

- Children can work out the value of the unknown number or symbol, using trial and improvement.

- Children can use inverses to work out the value of the unknown number or symbol, without fully appreciating the relationship.

- Children can explain how their understanding of inverses helped them work out the value of the unknown number or symbol.

Supporting the challenge

- Children write addition and subtraction number sentences only.

- Children use a number line to help with the addition and subtraction.

Extending the challenge

- Children write number sentences such as:

- Children remove one symbol and one number from each of their number sentences.

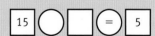

or

Can they identify more than one way of completing the calculation?

or

Colour the square

Using language associated with shape, movement and symmetry

Using and applying

Reasoning

Describe patterns and relationships involving shapes; make predictions and test these with examples

Communicating

Present solutions to puzzles and problems in an organised way; explain decisions, methods and results in pictorial, spoken or written form, using mathematical language

Maths content

Understanding shape

- Visualise common 2D shapes; identify shapes in different positions and orientations
- Identify reflective symmetry in patterns and 2D shapes

Key vocabulary

shape, pattern, balance, symmetry, symmetrical, line of symmetry

Resources

- NNS ITP: Symmetry
- RS30

For each child:

- RS31
- Thick crayons or felt-tipped pens
- Scissors

> How can you colour this square so that it is symmetrical?

Introducing the challenge

 Using the NNS ITP: Symmetry, highlight a number of squares to make a simple shape.

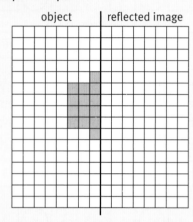

Explain to the children that you are going to colour more squares so that the shape is symmetrical. If the children are familiar with reflective symmetry, use appropriate language to explain what you are doing.

Repeat the above several times for other shapes, asking individual children to come to the board to complete the shape. If appropriate, change the line of symmetry from vertical to horizontal.

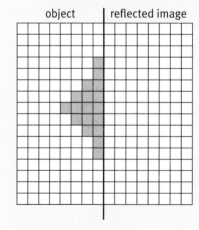
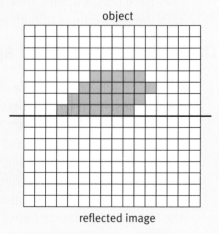

The challenge

 Provide each child with a copy of RS31, crayons or felt-tipped pens and scissors. Display the pattern from RS30 on the interactive whiteboard or as an enlarged poster. Children look at the patchwork squares and tell you what they can see in the pattern. Ask them to rotate the paper and look again.

Look at this patchwork. What do you notice? What else do you notice?

What shapes can you see?

Can you see a square? How many squares?

Can you see any diamonds?

Children then colour the squares carefully so that the design is symmetrical, using only four colours altogether.

Referring to the square on the interactive whiteboard or poster, colour one area within it. Children find other areas they could colour the same to balance the pattern.

Encourage the children to talk through how they will colour their first square. Remind them to use only four colours and to make it 'balance'. Do the children think they can colour all four squares so that each looks different? They must use the same four colours on each square.

 Give the children enough time to work on the challenge. Children can cut out the completed designs and twist and turn them to see if their designs are different.

Drawing out using and applying

 Arrange the children into groups and ask them to compare different squares to see where they are similar and where they are different. Encourage the children to turn the squares when looking for similarities.

 Children describe their finished squares, discussing similarities and differences between patterns.

How do they balance?

Which squares do you like? Why?

How are these two patterns alike? How are they different?

Assessing using and applying

- Children can make a symmetrical pattern.
- Children can make different symmetrical patterns, including twisting and turning their completed designs when looking for differences.
- Children can identify and explain similarities and differences between patterns, using the language of symmetry.

Supporting the challenge

- Children use two colours, rather than four, for each pattern.
- Start off one of the patterns for the children to copy.

Extending the challenge

- Children cut one coloured square into four quarters and fit these together to make a new square. They explore how many different designs they can made by rearranging pieces.
- Children take a square that has not been coloured and cut it into quarters. They colour each quarter with exactly the same design and explore the various square designs they can make.

Two-piece tangram

Exploring properties of regular and irregular two-dimensional shapes

Using and applying

Reasoning

Describe patterns and relationships involving shapes; make predictions and test these with examples

Communicating

Present solutions to puzzles and problems in an organised way; explain decisions, methods and results in pictorial, spoken or written form, using mathematical language

Maths content

Understanding shape

- Visualise common 2D shapes; identify shapes in different positions and orientations
- Make and describe shapes, referring to their properties

Key vocabulary

two-dimensional (2D) shape, regular, irregular, circle, triangle, square, rectangle, pentagon, octagon, hexagon, sided, sides, straight, curved, area

Resources

- Opaque bag with a collection of different 2D shapes
- Individual whiteboard and marker (for each child)
- Display copy of RS32

For each pair:
- RS32, scissors

For *Supporting the challenge*:
- RS33, glue and paper

For *Extending the challenge*:
- RS34

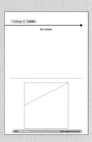

Can you name this shape?

Introducing the challenge

 Provide each child with an individual whiteboard and marker. Show the children the bag and tell them that it contains different 2D shapes. Explain that you are going to choose a shape from the bag secretly and describe it to them. Children draw the shape on their whiteboard. When they have done this, they turn over their whiteboard and wait until you tell them to show their drawing. Do this several times before inviting individual children to choose and describe a shape to the rest of the class.

What shape did I choose from the bag? How do you know?

What helped you finally decide which shape it was?

The challenge

 Arrange the children into pairs and provide each pair with a copy of RS32 and scissors. Using the display copy of the resource sheet, cut out the square and cut it into the two pieces. Each pair of children needs to cut their two pieces in the same way.

Ask a child to fit both pieces together to make a new shape. Discuss the new shape. Trace your finger around its perimeter to look at the sides. Count the number of sides the shape has. Can the children name the shape? Challenge the rest of the class to make one just the same. As a class, discuss whether you only allow whole sides touching.

 Working in pairs, children make shapes with a different number of sides: 3, 4, 5, 6 and 7 are all possible.

Can you find a three-sided shape?

What about a shape with four sides?

Can you find a different four-sided shape?

Children invent a suitable way of keeping track of the shapes they have made.

Drawing out using and applying

 Children show and discuss the shapes made.

Who found a shape with four sides? What about six sides?

Did anyone find a six-sided shape that looks different?

How are the shapes the same? How are they different?

Did anyone find a shape with seven sides? What about eight sides?

Encourage children to describe what the shapes look like to them in their own words. Give the shapes names.

Discuss the relationship between the areas of the different shapes made. Do the children appreciate that the area remains constant?

Assessing using and applying

- Children can make different shapes.
- Children can make, name and draw three-, four-, five-, six- and seven-sided shapes.
- Children can make different three-, four-, five-, six- and seven-sided shapes and draw them accurately.

Supporting the challenge

- Provide pairs of children with a copy of RS33, glue or sticky tape and a large sheet of paper. Children cut out the squares from the resource sheet, arrange each set of two shapes to make other shapes and stick these onto a large sheet of paper.
- Help the children identify reflections of shapes.

Extending the challenge

- RS34 provides a three-piece tangram.
- Challenge the children to cut up a square into four pieces to make their own puzzle.

Make it my way

Making and describing three-dimensional shapes and developing mental images of shape and position

Using and applying

Reasoning

Describe patterns and relationships involving shapes; make predictions and test these with examples

Communicating

Present solutions to puzzles in an organised way; explain decisions, methods and results in pictorial, spoken or written form, using mathematical language

Maths content

Understanding shape

- Visualise common 2D shapes and 3D solids; identify shapes in different positions and orientations; make and describe shapes, referring to their properties
- Follow and give instructions involving position

Key vocabulary

two-dimensional (2D), three-dimensional (3D), shape, position, describe, small, large, left, right, vertical, horizontal, side, circle, triangle, square, rectangle, cube

Resources

- Paper and pencil
- RS35
- RS36 (optional, for each child)
- Interlocking cubes of different colours (for each pair)

For *Supporting the challenge*:
- Building blocks

Introducing the challenge

Provide each child with paper and pencil. Tell the children that you are going to describe a picture to them (see RS35). This picture is made up of two circles, two, squares, two rectangles and two triangles: eight shapes altogether. As you describe the picture to them, children draw this on their sheet of paper. Make sure to use appropriate shape and positional language and differentiate each pair of identical shapes by size.

Draw a small circle on top of the large triangle.

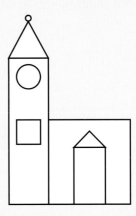

You may wish to invite children to ask you questions as you describe the picture.

Once you have completed the instructions, ask the children to show their completed drawings. Display the picture on RS35 on the interactive whiteboard or as an enlarged poster. Discuss with the children the similarities and differences between their picture and the one on the resource sheet. Also talk to them about what they found easy and difficult about the task and what words or phrases were important for them to draw an accurate picture.

The challenge

Arrange the children into pairs. Provide each child with a copy of RS36 and each pair with a supply of interlocking cubes of different colours. Introduce the challenge to the children. Pairs of children sit back to back. One child makes a model with a few interlocking cubes and instructs the other child how to make an identical model by describing their own model. Both children compare models to see how successful they were at giving and following instructions about three-dimensional shapes.

After the children have made their models, they make a representation of the two shapes on their resource sheet. They then swap roles and repeat the activity.

Drawing out using and applying

 Bring the class back together for a discussion. Pairs of children show the two models they made and tell the others how they described them, and why they were successful or unsuccessful in reproducing each other's models.

What shapes are easier to describe? Why?

What positions are hard to explain?

What other words and phrases could we use?

What are the difficulties in drawing three-dimensional shapes on paper?

How did you deal with these difficulties?

Is there any other way of doing it?

What can you tell me about three-dimensional shapes from pictures of them?

Assessing using and applying

- Children can describe and reproduce three-dimensional models to a certain degree of accuracy.
- Children can describe and reproduce three-dimensional models successfully, using a range of appropriate positional language.
- Children can draw recognisable representations of three-dimensional models on paper.

Supporting the challenge

- Tell the children to use no more than 10 interlocking cubes for their model.
- Children use building blocks to make their models.

Extending the challenge

- Children make more complex shapes or use other apparatus.
- Children make a model from each other's drawings.

Symmetrical patterns

Identifying reflective symmetry in patterns

Using and applying

Reasoning

Describe patterns and relationships involving shapes; make predictions and test these with examples

Communicating

Present solutions to puzzles in an organised way; explain decisions, methods and results in pictorial, spoken or written form, using mathematical language

Maths content

Understanding shape

- Identify reflective symmetry in patterns
- Follow and give instructions involving position

Key vocabulary

pattern, balance, symmetry, symmetrical, line of symmetry

Resources

- NNS ITP: Area or a square grid on a transparency and transparent coloured counters to display on an OHP

For each pair:

- RS37, supply of RS38
- 8 counters (one colour)

For *Supporting the challenge*:

- Pegboards and pegs of the same colour, elastic band

For *Extending the challenge*:

- Pile of different-coloured counters, coloured pencils

> **How many different patterns did you make, using eight counters?**

Introducing the challenge

 Using the NNS ITP: Area, set the grid size to 10 by 10. Click on the circle from the Shape Controls menu. Using the interactive whiteboard tools, draw a vertical line down the centre of the grid. Then click three squares on the grid to the left of the vertical line (that is, the line of symmetry) to make a simple shape.

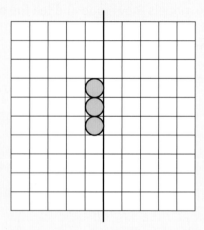

Remind the children of the line of symmetry and what it does. Explain that you are going to highlight three more squares on the grid so that the shape is symmetrical.

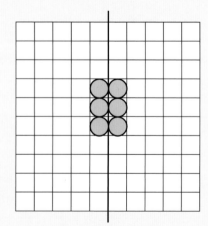

Repeat the above several times for other patterns, asking individual children to come to the board to complete the shape.

The challenge

 Arrange the children into pairs. Provide each pair with a copy of RS37 and eight counters. Also make sure that the children have access to a supply of RS38. Display RS37 on the interactive whiteboard or as an enlarged poster. Explain to the children that they need to take turns to place four counters on one side of the line of symmetry on their copy of RS37.

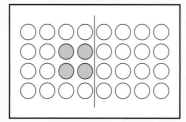

The other child then uses the four remaining counters to complete the pattern. Both children work together to record the completed pattern on one of the grids on RS38.

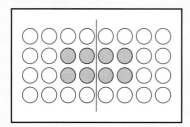

Children continue to take turns making and completing patterns, using only eight counters in total. Make sure the children realise that they need to check the patterns they have already made while working so that there are no duplicates.

How do you know that that counter belongs there?

Have you made this pattern before? How can you be sure?

How are these two patterns similar? How are they different?

Drawing out using and applying

 Using the NNS ITP: Area or RS37 displayed on the interactive whiteboard, individual pairs of children to show half of one of their patterns. The rest of the class suggests how to complete the pattern. Discuss with the children how they recorded their patterns and how they made sure that they did not produce any duplicates.

Assessing using and applying

- Children can make and complete simple symmetrical patterns involving whole shapes that are adjacent to the line of symmetry.

- Children can make and complete symmetrical patterns that are partly adjacent to the line of symmetry.

- Children can make and complete simple symmetrical patterns that are not adjacent to the line of symmetry.

Supporting the challenge

- Investigate making patterns, using four or six counters.
- Children make their patterns using pegboards and pegs of the same colour. They use an elastic band to create the vertical line of symmetry.

Extending the challenge

- Investigate making patterns, using 10, 12 ... counters.
- Children work with different-coloured counters.

From start to finish

Position, direction and movement

Using and applying

Representing

Identify and record the information needed to solve a puzzle; carry out the steps and check the solution in the context of the problem

Communicating

Present solutions to puzzles in an organised way; explain decisions, methods and results in pictorial, spoken or written form, using mathematical language

Maths content

Understanding shape

- Follow and give instructions involving position, direction and movement

- Recognise and use whole, half and quarter turns, both clockwise and anticlockwise; know that a right angle represents a quarter turn

Key vocabulary

position, direction, movement, turn, up, down, left, right

Resources

- NNS ITP: Area or a 5 by 5 grid on a transparency and two different-coloured transparent counters to display on an OHP
- RS39 (for each child)

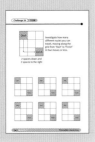

How many different routes can you travel from 'Start' to 'Finish' in four moves or less?

Introducing the challenge

 Using the NNS ITP: Area, set the grid size to 5 by 5. Choose the circle from the Shape Controls menu. Choose yellow from the Shape Colour menu and highlight a yellow circle in the top left-hand corner of the grid. Change the shape colour to green and highlight a green circle in the bottom right-hand corner of the grid. if working with an overhead projector, follow the instructions below by moving the counters.

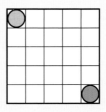

Tell the children that you want to move from the yellow circle to the green circle and that you can only move up or down and to the left or right – you cannot move diagonally. Use the interactive whiteboard tools to show one possible route, describing the route to the children as you go.

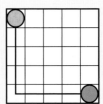

Down four squares, turn to the right and along four squares.

Children suggest other routes and show these on the interactive whiteboard. Make sure that the children describe both the turn and number of spaces moved.

The challenge

 Provide each child with a copy of RS39 and introduce the challenge to the class. Using the example at the top of the resource sheet, explain to the children the move from 'Start' to 'Finish', along the grid. For example: "One, two spaces down. One, two spaces to the right."

Accounting for all possibilities

Explain to the children how this was a two-step move: moving down was one step and moving to the right was another step. Children investigate how many different routes they can travel, moving along the grid from 'Start' to 'Finish' in four moves or less. Tell the children to use the grids at the bottom of the resource sheet to record their routes and write a brief description of the route.

 Give the children enough time to work independently on the challenge. Monitor children as they work, asking questions that will help them identify further routes or any duplicates they may have made. Also tell them to write a brief description of each route.

Describe this route to me.

Can you think of another route that is similar to this one?

How are you going to write about this route?

 Once the children cannot identify any more routes, arrange them into pairs to compare their routes.

Do you both have the same number of routes?

What is the same about your descriptions? What is different?

Can you see how this is another route?

Drawing out using and applying

 Display RS39 on the interactive whiteboard or as an enlarged poster. Provide an opportunity for children to describe one of their routes to the rest of the class. As they do so, draw the route on one of the grids at the bottom of the resource sheet. Occasionally, ask the children to describe the amount and direction of turn as they move through a route: for example, the route 'Right 2 spaces and down 2 spaces' has a quarter turn (or one right-angle turn) in a clockwise direction. Continue until all six routes are displayed.

Describe to us one of your routes.

How are these two routes similar? How are they different?

Who can describe to us one of their routes that we do not have here on the board?

How many different routes did you find?

Assessing using and applying
- Children can draw different routes from 'Start' to 'Finish'.
- Children can draw and write about different routes from 'Start' to 'Finish'.
- Children can draw and write about all six different routes possible from 'Start' to 'Finish'.

Supporting the challenge
- Children work in pairs.
- Show the children a three-stage and a four-stage move as further examples.

Extending the challenge
- How do the directions change if 'Start' and 'Finish' are in different places?

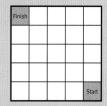

- Investigate different routes from 'Start' to 'Finish' on this grid.

Turning

Recognising whole, half and quarter turns, both clockwise and anticlockwise

Using and applying

Reasoning

Describe patterns and relationships involving shapes; make predictions and test these with examples

Communicating

Present solutions to puzzles in an organised way; explain decisions, methods and results in pictorial, spoken or written form, using mathematical language

Maths content

Understanding shape

- Recognise and use whole, half and quarter turns, both clockwise and anticlockwise; know that a right angle represents a quarter turn
- Follow and give instructions involving position, direction and movement

Key vocabulary

turn, whole, half, quarter, clockwise, anticlockwise, position, direction, movement, left, right, right angle, point, line

Resources

- Everyday classroom objects that turn: scissors, clock, toy car
- RS40
- RS41 (for each child)

What do you notice about some of the turns between different points?

Introducing the challenge

 Ask the children to suggest objects that turn. If readily available, use everyday classroom objects: for example, scissors, doors, hands on a clock, wheels on toy cars.

Can you tell me something that turns?

How is a door turning similar to the minute hand on a clock turning?

How is a door turning different from a hand on a clock turning?

The challenge

 Display RS40 on the interactive whiteboard or as an enlarged poster and give each child a copy of RS41. Discuss with the children how a wheel is an example of something that turns about a point.

Briefly revise children's understanding of whole, half and quarter turns, both clockwise and anticlockwise, and how one right angle represents a quarter turn. Use the wheel on RS40 and RS41 to assist with the explanations.

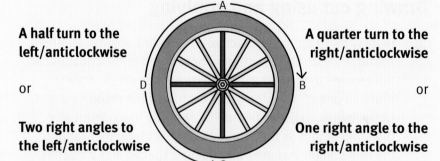

A half turn to the left/anticlockwise

or

Two right angles to the left/anticlockwise

A quarter turn to the right/anticlockwise

or

One right angle to the right/anticlockwise

Can you describe to me how you would get from Point B to Point C?

Is there another way?

What about from Point B to Point D?

Briefly introduce the challenge to the children and ask them to investigate different turns that go between two points. Discuss how they might record the different turns. Also make sure they realise that there are different ways of explaining any one turn.

From A to B:

- **quarter turn clockwise ($\frac{1}{4}$ CW)**
- **quarter turn to the right ($\frac{1}{4}$ to the R)**
- **three-quarters turn anticlockwise ($\frac{3}{4}$ ACW)**
- **three-quarters turn to the left ($\frac{3}{4}$ to the L)**
- **1 right angle to the right**
- **1 right angle in clockwise direction**
- **3 right angles to the left**
- **3 right angles in an anticlockwise direction**

Drawing out using and applying

Display RS40 again and invite children to share with the rest of the class the different turns that go between two points. Show these on the wheel on the resource sheet. Ask other children to explain the same move using different terminology. Discuss which move(s) are the most direct. For example, From A to B: a quarter turn clockwise or a quarter turn to the right is more direct than a three-quarters turn anticlockwise or a three-quarters turn to the left.

Can you describe the turn from Point B to Point C?

Who can explain this turn in a different way?

Which is the most direct turn?

What about from Point A to Point C?

How would you describe a turn from A and back to A again?

What about from C to C?

Assessing using and applying

- Children can identify different turns in a haphazard way.
- Children can work systematically to identify all the different turns possible.
- Children can identify relationships between different turns: for example, turns A to D and D to A; all the half turns. They can also recognise the most direct turn between two points.

Supporting the challenge

- Help the children work systematically to identify the different turns possible, always starting with the clockwise turn, then moving on to the anticlockwise turn.
- Suggest abbreviations for recording the different turns. For example: From A to B: $\frac{1}{4}$ CW or $\frac{3}{4}$ ACW.

Extending the challenge

- Children describe each turn as the least amount of turn possible. For example: From A to B: "A quarter turn clockwise" or "A quarter turn to the right" rather than "A three-quarters turn anticlockwise" or "A three-quarters turn to the left".
- Children describe the turns as right angles. For example: From A to B: "1 right angle to the right" or "3 right angles to the left".

Making a ruler

Making and evaluating a simple measuring instrument

Using and applying

Enquiring

Follow a line of enquiry; answer questions by choosing and using suitable equipment and selecting, organising and presenting information in simple diagrams

Communicating

Present solutions to problems in an organised way; explain decisions, methods and results in pictorial, spoken or written form, using mathematical language

Maths content

Measuring

- Estimate, compare and measure lengths, choosing and using standard units and suitable measuring instruments
- Read the numbered divisions on a scale and interpret the divisions between them

Key vocabulary

length, unit, equipment, measure, measurement

Resources

For each child:

- RS42
- Strips of paper of different lengths
- Interlocking cubes

For *Extending the challenge*:

- 1 cm cubes
- Standard ruler

Which rulers work? Why?

Introducing the challenge

 Join enough cubes together in a rod to reach across the table. As you join the cubes, work with the class to estimate the length of the table in cubes. Encourage the children to picture the cubes in their minds and adjust their estimates as the rod is nearing completion.

How many cubes long do you think this table is?

Who thinks it is longer? Shorter?

Who wants to change their estimate?

Do you think it will be longer or shorter than you originally estimated?

Choose something shorter than the completed rod of cubes: for example, the width of the table. Children estimate this distance in cubes.

Use the rod of cubes as a 'ruler' to measure the new item. Discuss how far along the ruler the new item reaches: for example, halfway? Nearly halfway?

Repeat for other objects.

The challenge

 Provide each child with a copy of RS42, strips of paper of different lengths and a supply of interlocking cubes. Tell the children that the challenge is to make their own 'cube ruler', using paper. The rulers must 'work', and the test will be to use the finished rulers to measure the top of the table to see if they all agree.

 Give the children a maximum of three minutes to discuss in pairs how they will tackle the challenge.

 Briefly bring the class back together and share their ideas.

How might you go about making your ruler?

What are you going to use to help you?

What are you going to include on your ruler?

All the children make their own ruler, but work in pairs to share methods of solving the problem. The rulers will be different lengths. Some children may choose to number them, while others may not.

Drawing out using and applying

 Children explain how they solved the problem and justify their decisions.

Gather together a selection of the children's rulers, lay them side by side and compare them. Discuss what the children notice. What do they understand by rulers that 'work'?

Use the rulers to measure the length and width of the table. Record these measurements on the board and discuss the results.

Why do you think that these two rulers give different readings?

Why did these two rulers give similar readings?

What makes this ruler 'work'?

How could this ruler be improved?

Assessing using and applying

- Children can make a ruler using paper.
- Children can make a successful ruler using paper and use it to accurately measure the top of the table.
- Children can critically evaluate their ruler, commenting on its strengths and limitations and suggesting ideas as to how it might be improved.

Supporting the challenge

- Children work in pairs to create one ruler.
- Suggest the children make their own 'cube ruler' the length of the desk and use this to help them make a 'paper ruler'.

Extending the challenge

- Use the ruler to find the length of something longer than the ruler itself. How can we solve the problem without making the ruler longer?
- Compare 1 cm cubes with a standard ruler. Measure the same item, using both methods.

Puffed and popped

Using non-standard and standard measures and simple measuring equipment for weight

Using and applying

Enquiring
Follow a line of enquiry; answer questions by choosing and using suitable equipment and selecting, organising and presenting information in lists, tables and simple diagrams

Communicating
Present solutions to puzzles and problems in an organised way; explain decisions, methods and results in pictorial, spoken or written form, using mathematical language

Maths content

Measuring
- Estimate, compare and measure weights, choosing and using standard units and suitable measuring instruments

Handling data
- Answer a question by collecting and recording data

Key vocabulary
weight, volume, amount, more, less, the same, balances, light, lighter, lightest, heavy, heavier, heaviest

Resources
- Set of balances, different classroom objects
- RS43 (optional, for each child)

For each group:
- Rice or cereal, yoghurt pots
- Balances and weights: 50 g and 100 g

Which is heavier, an uncooked grain or a puffed grain, or are they the same?

Introducing the challenge

 Show the children a set of balances and remind them of its purpose. Briefly discuss where they may have seen a set of balances before and what it might be used for.

What do we call this?

What is it used for?

Where have you seen these before?

What do people do with them?

How do they use them?

Why are they useful?

Choose two classroom objects and place each one on either side of the balances. Discuss the results with the children. Encourage them to use appropriate mathematical language: for example, 'more', 'less', 'balances', 'lighter' and 'heavier'. Repeat several times.

Place an item on one side of the balances and ask a child to come and chose an item that they think will be heavier (or lighter) than the object already on the balances. Again, repeat several times.

Next, place one of the standard weights (for example, 100 g) on one side of the balances and a classroom object on the other side. Tell the children the weight of the standard weight and ask them to make a true statement. Repeat several times, choosing combinations of standard weights if appropriate.

The challenge

 Arrange the children into groups and provide each group with the necessary resources and each child with a copy of RS43. Introduce the challenge to the class. Tell the children to use the yoghurt pots to measure out quantities of rice or cereal that they predict will weigh the same. They then use the balance to see what happens and adjust the quantities until the pots balance. They record their findings.

How much rice do you think you will need so that it weighs the same as this amount of cereal?

> Children take the same amount of rice or cereal and see if they weigh the same.

How many rice grains do you think a pot contains?

How could you find out

> Children use a weight, for example, 50 g, and find out how much rice or cereal they get for that weight. Encourage them to estimate first. They record their results on a chart, showing the weight, then the number of potfuls of each kind. They can draw the chart on the back of RS43.

How are you going to record your results?

Drawing out using and applying

 Discuss the challenge with the children. If groups used different types of cereals, invite them to explain their results to the rest of the class. How were the results similar? How were they different?

Which is heavier: a grain of rice or a flake of cereal, or are they the same?

How can you tell?

Were you surprised by any of your results?

Assessing using and applying
- Children step back from the challenge and contribute little to the group.
- Children can contribute to the group and organise and present their results in a clear and logical way.
- Children can make predictions based on previous observations.

Supporting the challenge
- Arrange the children into friendship rather than ability groups, thus making it easier for all children to make a contribution to their group.
- Help the children record their results.

Extending the challenge
- How much uncooked grain would give you enough for a breakfast bowl of puffed grain?
- Children find out how much a potful of each kind of grain weighs.

Which is the larger?

Making and testing estimates of capacity

Using and applying

Enquiring

Follow a line of enquiry; answer questions by choosing and using suitable equipment and selecting, organising and presenting information in lists and simple diagrams

Communicating

Present solutions to puzzles and problems in an organised way; explain decisions, methods and results in pictorial, spoken or written form, using mathematical language

Maths content

Measuring

- Estimate, compare and measure capacities
- Choose and use suitable units and measuring instruments

Key vocabulary

most, more, holds, capacity, smaller, smallest, larger, largest

Resources

For each pair:
- RS44 (optional)
- Small boxes or plastic containers with wide necks
- Selection of counting materials: Compare Bears, counters, beads, interlocking cubes

For *Extending the challenge*:
- 10 conkers, card, scissors, sticky tape and other construction material

Which do you think holds the most?

Introducing the challenge

 Hold up two boxes of similar sizes, but different shapes, and discuss as a class which the children think is the larger. Discuss what they mean by 'larger'. Draw their attention to the inside of each box. Which do they think holds the most? Ask for ideas about how to find out. Try out their ideas to discover which box is the larger. Discuss the results.

Which of these boxes do you think is the larger? Why do you say that?

What does 'larger' mean?

Does it mean anything else?

Which of these boxes do you think holds the most? Why do you say that?

How can we find out which box holds more?

How can we find out which is the larger, using beads?

Tell me a true statement about these two boxes.

The challenge

 Arrange the children into pairs and provide each pair with the resources. Each pair of children chooses three containers that look interesting to compare and compares how much the containers hold.

Encourage the children to devise appropriate ways of recording their findings on their copy of RS44.

 When pairs have compared their three containers, arrange them into small groups and ask each pair to share their methods and findings with the other children in the group. Can the group then order all the containers from smallest to largest?

Drawing out using and applying

 Bring the class back together again and share their methods and results.

Were there any surprises?

Discuss any different measuring units they used. Encourage the children to respond to questions such as:

What would happen if you used conkers instead of beads?

Share the different ways of recording the children have used and the effectiveness of each.

Assessing using and applying

- Children can compare three containers, saying which is the largest.
- Children can compare and order a set of containers, explaining and justifying their methods and findings.
- Children can make comparisons between both similar and vastly different containers and explain how capacities do not change regardless of the measuring unit being used.

Supporting the challenge

- Provide the children with three vastly different-sized containers.
- Remind the children of the importance of using uniform non-standard measuring equipment.

Extending the challenge

- Can the group order all the containers from smallest to greatest capacity?
- Can they find (or make) a box that will hold exactly 10 conkers?

Just a minute!

Using counting and a simple timer to measure duration of time

Using and applying

Representing

Identify and record the information needed to solve a problem; carry out the steps and check the solution in the context of the problem

Communicating

Present solutions to problems in an organised way; explain decisions, methods and results in pictorial, spoken or written form, using mathematical language

Maths content

Measuring

- Estimate, compare and measure time, using standard units and suitable measuring instruments
- Use units of time; identify time intervals

Key vocabulary

time, how many?, minute

Resources

- NNS ITP: Tell the time
- One-minute timer or stopwatch
- RS45 (optional, for each child)

For each pair:

- One-minute timer, other materials to help measure time: interlocking cubes, skipping rope, beads and laces

For *Extending the challenge*:

- One-minute tocker, plastic bottle, sand and construction material

What are some of the things that you can do in one minute?

Introducing the challenge

 Using the NNS ITP: Tell the time, show both the analogue and 12-hour digital clocks. Show the seconds hand on the analogue clock and the seconds on the digital clock. Set the clocks to the current time and start the clocks.

Tell the children that they are going to see exactly how long one minute lasts. Wait until the start of a new minute: for example, 10:12:00. If you have a large class clock with a second hand, you could get the children to watch this instead.

Explain to the children that they are going to do this again, but this time, they have to estimate how long they think one minute is. Hide the NNS ITP: Tell the time (or instruct the children not to look at the class clock) and show the children the one-minute timer or stopwatch. Tell the children that you will say 'Start' and that they have to estimate how long they think one minute is. When they think they have reached one minute, they put their hand up in the air. When the minute is up, you say 'One minute!'

Briefly discuss with the children things they think that last one minute or that they can do in one minute.

The challenge

 Arrange the children into pairs. Provide each child with a copy of RS45 and each pair with a one-minute timer and a selection of materials to help measure time. Talk through the activities on the resource sheet. Children do the activities for one minute. Child A is the timekeeper, and Child B does the activities. Children then swap roles.

Children suggest other things they could try doing in one minute.

 After they have done a few activities, ask the children to estimate beforehand how many repetitions they think they will be able to do. Remind them to record their work. You could challenge them to repeat an activity and see if they can do it again more times in one minute.

How many times do you think you will be able to skip in one minute?

How close was your estimate?

How are you going to record this?

Drawing out using and applying

 Have a class discussion about the results of the children's work. Compare the results of the activities that several children did and discuss the differences.

Why did the number of cubes fitted together vary from child to child?

What about writing your name or counting?

Why do these activities vary?

Does the minute change each time? What varies?

Assessing using and applying

- Children can write a list of different things they can do in one minute.
- Children can estimate how many repetitions they think they are able to do in one minute.
- Children appreciate that time does not vary, rather it is the length of time it takes to complete a task that varies.

Supporting the challenge

- Suggest other activities the children could do in one minute: for example, skip, thread beads onto a lace or walk from one side of the room to the other.
- If undergoing activities for which there will be no physical evidence of how many were completed in one minute (for example, skipping), arrange the children into groups of three: the child doing the task, a timekeeper and someone to tally the number of times the activity is completed.

Extending the challenge

- Use a tocker instead and count how many times it tocks while you put on your shoes, write your name once, thread 10 beads ...
- Make a minute timer with a plastic bottle and some sand.

Birthday month

Using standard units of time

Using and applying

Enquiring

Follow a line of enquiry; answer questions by choosing and using suitable equipment and selecting, organising and presenting information in lists, tables and simple diagrams

Communicating

Present solutions to problems in an organised way; explain decisions, methods and results in pictorial, spoken or written form, using mathematical language

Maths content

Measuring

• Use units of time and know the relationships between them

Handling data

• Answer a question by collecting and recording data in tables

Key vocabulary

time, day, week, month, year, calendar, birthday

Resources

• Large current year's calendar
• Class register (to refer to children's birthdays), RS46 (for each child), calendars and diaries for years other than the current year

For *Extending the challenge*:

• Card, scissors, sticky tape and other construction material

How did you make your birthday month calendar?

Introducing the challenge

 Display the current year's calendar and discuss it with the children. Ask questions similar to the following that require the children to read and interpret the calendar.

What is this?

What is it used for?

What does it tell us?

Why is it useful?

Let's look at the page for this month. What will be the date next Wednesday?

On what day of the week is the 21st?

How many Mondays are there this month?

What was the date last Friday?

Give the calendar to a child and ask them to make a true statement about the calendar. Repeat with other children.

Still referring to the current year's calendar, turn to the month in which you were born. Find your date of birth and tell the children on which day of the week your birthday fell/falls this year.

Ask the children if anyone knows on which day of the week their birthday fell/falls this year. For those children who do not know, tell them on which day of the week their birthday was/is this year.

The challenge

 Provide each child with a copy of RS46 and assorted calendars and diaries. Children use the calendars and diaries to find out the day their birthday fell/falls this year. They record the day their birthday fell/falls this year on the resource sheet and also the month in which they were born. Collect in the calendars and diaries before children move on to the next part of the challenge

Children make a calendar for their birthday month. They need to find out the number of days in that month.

How are you going to start to make a calendar for your birthday month?

 When two children with the same birthday month have finished, pair them up and ask them to compare their calendars.

 When all the children in the class have prepared a birthday calendar, make a class calendar by putting all the months together. Children can use it to record class and school events.

Drawing out using and applying

 Children share and discuss their completed calendars.

How many days are in your birthday month?

What day of the week is the first day of the month?

Who has a birthday in September? Did you compare your calendars? Were they the same?

Do any of you share the same birthday? How did your calendars compare with each other?

Look at the patterns in the calendar, pointing out similarities and differences between the months. Look at calendars for different years, and how birthdays fall on different days.

Assessing using and applying
- Children can make a calendar for their birthday month after some assistance.
- Children can successfully make a calendar for their birthday month.
- Children can identify and comment upon similarities and differences between months, and why birthdays fall on different days in different years.

Supporting the challenge
- Allow the children access to calendars or diaries.
- Write the child's date of birth in the correct position on the resource sheet.

Extending the challenge
- Children make their own individual calendar of the whole year and assemble it themselves.
- Look at the patterns on the calendar by colouring in, for example, all the Mondays in one month. Look at the dates and investigate that sequence of numbers.

Toy sort

Sorting objects using lists, tables and diagrams

Using and applying

Enquiring

Follow a line of enquiry; answer questions by choosing and using suitable equipment and selecting, organising and presenting information in lists, tables and simple diagrams

Communicating

Present solutions to problems in an organised way; explain decisions, methods and results in pictorial, spoken or written form, using mathematical language

Maths content

Handling data

- Answer a question by collecting and recording data in lists and tables; represent the data as block graphs or pictograms to show results
- Use lists, tables and diagrams to sort objects; explain choices using appropriate language, including 'not'

Key vocabulary

sort, organise, list, table, diagram

Resources

- RS47 (optional, for each child)
- Large collection of different toys (for each group)

For *Extending the challenge*:

- RS48, RS49 or RS50

How else can you sort the toys?

Introducing the challenge

 Tell the children that you want to find out what is the class's overall favourite toy. Individual children name their favourite toy. Accept a broad range of toys, including sports equipment and computer games. Write a list of these on the board. Through discussion with the children, narrow down the list to the five or six most popular toys.

football	**doll**	**bicycle**	**puzzle**
board game	**skateboard**	**skipping rope**	**cooking set**

remote-control car

Ask the children how they could find out what is Year 2's favourite toy. Lead the discussion towards making a table of the top five or six toys and everyone in the class voting for one of these.

Toy	Number
football	
puzzle	
skateboard	
doll	
bicycle	

Once the table is complete, ask questions that require the children to interpret the results in the table.

What is the most popular toy?

What is the second most popular toy?

How many more of you voted for the skateboard?

Which toy got eight votes?

Which toys got more votes than the bicycle?

What can you say about the types of toy that are the most popular? What types of toy are they?

The challenge

 Arrange the children into groups and provide each child with a copy of RS47 if appropriate and each group with a large selection of different toys. Do not spend time discussing how to sort the toys; rather, let children come up with their own criteria. Make sure, however, that the children realise that they need to show how they have sorted their toys, using pictures or diagrams, and to record this on the back of their copy of the resource sheet.

How are you going to sort these toys?

How are you going to show this on the resource sheet?

How else could you sort the toys?

Is there another way?

How are you going to keep a record of this so that other people can read and understand it?

Drawing out using and applying

 Bring the whole class back together and provide an opportunity for individual children to explain to the rest of the class how they sorted their toys. Ask children from the same table to suggest other ways of sorting the toys. Direct the discussion towards how children presented the outcomes of their sorting.

Jakira, tell us one way you sorted your toys.

Show us what you wrote down.

Did anyone on Jakira's table sort the toys in the same way? What did you write down?

Did anyone on Jakira's table sort the toys a different way?

Who sorted their toys in a different way? How did you record this?

How many different ways did people find to sort their toys? What are they?

Assessing using and applying

- Children can sort the toys according to a given criterion.
- Children can sort the toys according to their own criterion.
- Children can sort the toys according to different criteria.

Supporting the challenge

- Children sort the toys into groups according to a given criterion: for example, by size, purpose, indoor or outdoor.
- With the support of an adult, children take it in turns to sort the toys without explaining the criterion. In the other group, members work out what the rules of sorting are.

Extending the challenge

- Children sort the toys into two, three or four different groups.
- Provide the children with a copy of RS48, RS49 or RS50 and ask them to represent their results in a pictogram or block graph.

Brothers and sisters

Representing data in a pictogram

Using and applying

Enquiring

Follow a line of enquiry; answer questions by choosing and using suitable equipment and selecting, organising and presenting information in lists, tables and simple diagrams

Communicating

Present solutions to problems in an organised way; explain decisions, methods and results in pictorial, spoken or written form, using mathematical language

Maths content

Handling data

- Answer a question by collecting and recording data in lists and tables; represent the data as pictograms to show results
- Use lists, tables and diagrams to sort objects; explain choices using appropriate language, including 'not'

Key vocabulary

collect, organise, present, record, pictogram, symbol

Resources

For each group:

- Individual whiteboard and marker (optional)
- Enlarged copy each of RS48, RS49, RS51
- Ruler
- Coloured pencils

How many brothers or sisters does each child in our class have?

Introducing the challenge

 Arrange the children into groups and explain to the class that they are going to work together in groups to answer the following question:

How many brothers or sisters does each child in our class have?

 Ask the children to briefly discuss as a group what the question means and how they might go about finding out the answer. Make sure that the children interpret the question correctly and realise that if their family consists of two children, they have one brother or sister; likewise, if there are three children in their family, they have two brothers or sisters. Families are often complicated, and there will be step- and half-brothers and sisters to consider. Children choose which members of their family to include.

 After enough time, bring the class back and ask groups to offer their suggestions. Briefly discuss these with the class.

Discuss with the class how impractical it is for each child to go around the class asking every other child how many brothers and sisters they have. Ask the class for suggestions to overcome this problem. Feed in the idea that one person from each group could be responsible for asking the children in the other groups how many brothers and sisters they have. To make the data collection even more efficient, you may want to suggest that each group writes down their results on one whiteboard. You can then display it to the class.

Rachel	2	Amil	3
Prabha	1	Nathan	1
Craig	0		

The challenge

 Provide each group with an enlarged copy of RS48, RS49 and RS51, a ruler and coloured pencils. Referring to RS51, explain the task. Make sure that the children realise what a pictogram is and what 'collect', 'organise' and 'present' information means. (You may want to provide examples.) Children discuss the first three questions as a group before embarking on the challenge.

Discuss how children can use the squared paper on RS48 or the blank pictogram on RS49 to present their data. Discuss with individual groups how to label the vertical axis on RS49.

More than 3 brothers or sisters	
3 brothers or sisters	
2 brothers or sisters	
1 brother or sister	
0 brothers or sisters	

 Give groups enough time to work on the challenge. Monitor their progress as they work, asking questions where appropriate.

Why have you decided to organise your information in this way?

How might a table help you organise your information?

What symbol are you going to use in your pictogram?

What labels are you going to write on your pictogram?

Drawing out using and applying

 When groups have completed their pictogram, draw their attention to the last two questions on RS51.

 Bring the whole class back together again. Each group presents their completed pictogram and talks about their responses to the five questions on RS51. Discuss the similarities and differences between each group's method of collecting, organising and presenting the data. Did each group come up with the same results?

Assessing using and applying

- Children can collect, organise and present data with assistance or using a pre-populated pictogram.
- Children can collect, organise and present data in a pictogram they have constructed themselves.
- Children can collect, organise and present data in a pictogram and reflect on its strengths and weaknesses.

Supporting the challenge

- Help children organise their results into a table.

Number of brothers and sisters	Number of children
0	
1	
2	
3	
4	
5	

- Provide children with a copy of RS49, with the vertical axis labelled as shown.

Extending the challenge

- Suggest groups draw their pictogram where the symbol represents two children. How are they going to show one child?
- Children write three statements they can infer from the data they have presented in their pictogram.

Birthday months

Representing the same data in different block graphs

Using and applying

Enquiring

Follow a line of enquiry; answer questions by choosing and using suitable equipment and selecting, organising and presenting information in lists, tables and simple diagrams

Communicating

Present solutions to problems in an organised way; explain decisions, methods and results in pictorial, spoken or written form, using mathematical language

Maths content

Handling data

- Answer a question by collecting and recording data in lists and tables; represent the data as block graphs to show results
- Use lists, tables and diagrams to sort objects; explain choices using appropriate language

Key vocabulary

collect, organise, present, record, block graph, similarities, differences, strengths, weaknesses

Resources

- Enlarged copy of RS52
- Scissors, glue

For each pair:

- RS53 {enlarged}, coloured pencils

- RS50

For *Extending the challenge*:

- RS48

In which month were most children in the class born?

Introducing the challenge

 In advance, cut out and glue together the monthly calendar from the enlarged copy of RS52. Stick this up on the board.

January	February	March	April	

Explain to the children that you want to find out in which month most of the children in the class were born. Children suggest how to go about finding this out, using the monthly calendar on display.

How are we going to use this calendar to find out in which month of the year most of you were born?

Who was born in January?

How many children were born in January?

Continue until the calendar is complete.

January	February	March	April	
3	**2**	**4**	**5**	

Briefly discuss the results with the children.

In which months of the year were three children born?

How many children were born in March?

Were more children born in September or in July? How many more?

The challenge

 Arrange the children into pairs and provide each pair with an enlarged copy of RS53 and coloured pencils. Explain that they are to use the information presented in the monthly calendar on the board to complete the block graph. Children have 10 minutes in which to do this.

 Once the time is up, pairs display and comment upon their completed block graph. Discuss similarities and differences between different block graphs.

Next, give each pair a copy of RS50. Discuss the similarities and differences between the block graphs on the two resource sheets. Explain to the class that you want them to show the same results, but on a different block graph. Pairs discuss how they are going to go about completing the block graph. Do not discuss this part of the challenge any further with the class.

 As pairs work on the second block graph, monitor their progress. If necessary, ask questions that assist them in completing the graph.

How many columns does this graph have?

How might you change the labelling at the bottom of each column to fit in all 12 months?

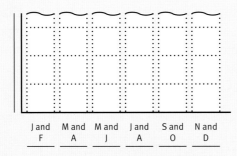

Drawing out using and applying

 Once pairs have completed their second graph, they display and comment upon their block graph. Discuss similarities and differences between the different block graphs made, using RS50. Also discuss the similarities and differences and strengths and weaknesses between the two different types of block graphs made, using RS53 and RS50.

Look at the two block graphs you have drawn. How are they similar? How are they different?

What is good about the way the information is presented on the first block graph? What about the second?

Can you use the second block graph to answer our question: 'In which month were most children in our class born?' Why not?

Assessing using and applying

- Children can complete both types of block graphs with encouragement.
- Children can recognise similarities and differences between the two different types of block graphs.
- Children can recognise and talk about the similarities and differences and strengths and weaknesses between the two different types of block graphs.

Supporting the challenge

- Help children complete both axes on each of the block graphs: in particular, the axes on the second block graph.
- Referring to the monthly calendar on the board, assist the children in grouping the months.

Extending the challenge

- Children write three statements they can infer from the data they have presented in their pictogram.
- Children collect data on the day of the month children in the class were born on and present it in a suitable block graph.

Dice totals

Choosing the most appropriate data-handling methods to answer a question

Using and applying

Enquiring

Follow a line of enquiry; answer questions by choosing and using suitable equipment and selecting, organising and presenting information in lists, tables and simple diagrams

Communicating

Present solutions to problems in an organised way; explain decisions, methods and results in pictorial, spoken or written form, using mathematical language and number sentences

Maths content

Knowing and using number facts

- Derive and recall all addition facts for each number to at least 10

Handling data

- Answer a question by collecting and recording data in lists and tables; represent the data as block graphs or pictograms to show results

Key vocabulary

collect, organise, present, record, list, table, pictogram, block graph, add, sum, total, plus

Resources

- Two large 1–6 dice
- RS48, RS54, two 1–6 dice, pencils (for each pair)

For *Supporting the challenge*:

- Interlocking cubes

For *Extending the challenge*:

- Two 0–9 dice

Roll two 1–6 dice and add the two numbers together. What is the most likely total?

Introducing the challenge

 Pose the following question to the class:

When you roll two 1–6 dice and add the two numbers together, what is the most likely total you will get?

Briefly discuss the question with the children, using two large 1–6 dice to demonstrate. Make sure that the children are clear about what the question is asking.

The challenge

 Arrange the children into pairs and provide each pair with a copy each of RS48 and RS54, two 1–6 dice and coloured pencils.

Give the children enough time to work on the challenge. Do not direct them towards one form of data organisation (for example, a table) or presentation (for example, a block graph or pictogram), but let them come to these decisions for themselves.

How are you going to keep a record of the different totals?

How are you now going to organise these results?

How are you going to present your final results? Why have you decided to show them like this?

Drawing out using and applying

 Bring the class back together again and provide an opportunity for pairs of children to report back their results to the rest of the class. Focus the discussion on the different ways children recorded, organised and presented their results, commenting on the strengths and weaknesses of each.

How did you keep a record of your results?

How did you organise your results? Did you use a table?

Who organised their results in a different way?

Collecting, organising and presenting data in different ways

How did you present your final results?

Who presented them in a different way?

What was good about the way you recorded, organised and presented your results?

What might you do differently if you had to do this activity again?

So, when you roll two 1–6 dice and add the numbers together, what is the most likely total you will get?

Assessing using and applying

- Children can record, organise and present their results in an ad hoc fashion.
- Children can record their results and subsequently organise them.
- Children can record their results systematically, organise them into a table and present them in a chart.

Supporting the challenge

- Help the children organise their results into a table.

Total	Number of rolls
2	
3	
4	
5	
6	
7	
8	
9	
10	
11	
12	

- Children use interlocking cubes to make a visual chart of their results.

Extending the challenge

- Suggest the children present their results in a block graph or bar chart.
- Children predict, then test what they think would be the most common total, using two 0–9 dice.